AI SALES ENGINE

How CEOs Architect, Deploy & Scale AI-Powered Revenue Systems

By Doug Foley

International Bestselling Author of Breakout Blueprint
&
CEO of FoleyMedia.ai | Host of 9 Figure Exits

ISBN:

979-8-9949273-1-1 Hardcover

979-8-9949273-3-5 Paperback

979-8-9949273-0-4 Kindle

979-8-9949273-2-8 Audible

Published by Foley Media, LLC

Printed in the United States of America

AI SALES ENGINE

How CEOs Architect, Deploy & Scale AI-Powered Revenue Systems

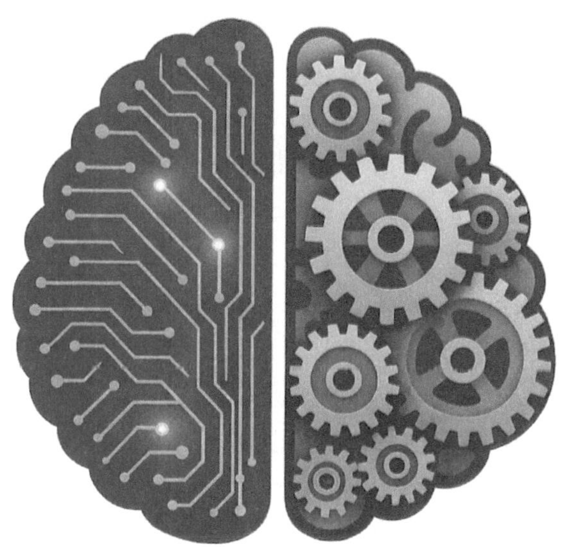

DOUG FOLEY

INTERNATIONAL BESTSELLING AUTHOR OF
BREAKOUT BLUEPRINT

Acknowledgments

To my friends: thanks for pretending to understand when I said, "it's just a small AI experiment," and then disappeared into a weekend of whiteboards, dashboards, and questionable snack choices. Your patience, laughs, and reality checks kept me human while I tried to make machines a little smarter.

To my clients: you trusted me to build engines, not just decks. You let me test, learn, and ship in the real world - where results count and excuses don't. This book exists because you chose execution over theater. I'm grateful for your partnership and the high bar you set.

To my mentors: thank you for lending me your playbooks - and for making me write my own. You taught me that strategy is a promise and shipping is the proof. Every clean framework in these pages stands on your shoulders.

To my teams - past and present: the builders, operators, analysts, RevOps whisperers, editors, designers, SDRs, PMs, and the folks who quietly make impossible things look routine. You turned sketches into systems, slideware into revenue motions, and "one more tweak" into a launch. Thanks for wrangling CRMs, taming data gremlins, humoring my late-night Slack pings, and tolerating file names like v2-final-FINAL-actually-final.ai. None of this ships without you.

To my family - especially my wife and kids: you are the best part of my life's operating system. You taught me the hardest lesson in business: build something great and be home for the good stuff. This book is as much about balance as it is about growth because of you.

To the readers and builders: may these pages save you time, earn your trust, and help you turn AI from a buzzword into a bottom-line advantage. If you find a typo, it's mine. If you find a breakthrough, it's ours.

PREFACE

"Dude, this is @#$% crazy...it's going to change our business!" That was the reaction I got after showing a client how to use AI to solve one of their bigger process bottlenecks...

All of this started about three months prior when we were at a charity golf tournament and the CEO asked me, "I see you doing all this crazy stuff with AI, but as a wholesaler who sells B2B I just don't get how it can help us?"

While we were driving to our next shots, I asked the CEO, what's one of the biggest challenges you have with your sales team? He said, "consistent outreach. I've got a young team and we don't do the same thing or they may not understand how to position us."

I opened up ChatGPT, pulled up the Custom GTM GPT I had built, and told him, "in a second I'm going to hit this little mic button and I want you to ask Chat how to solve that problem." While we went and made birdie, one of 14 on the day to a -14 finish, Chat gave a perfect answer; a full follow sequence, call scripts, voicemail ideas, and email scripts.

This is probably one of the most common things I hear from B2B businesses. Sure AI is cool at all the marketing things, but few realize how to maximize its true power and unlock ultimate efficiency across their entire sales and marketing teams.

Three weeks later I was in their boardroom troubleshooting their biggest bottleneck in their quoting process. Their business is driven by Government RFPs, which means constantly downloading faxed copies of PDFs then recreating pricing tables to prepare the quote for submissions. On this particular day, they actually had a table of all the items they needed to quote, which they referred to as a kit. I went back to the warehouse, took a few photos of all the items on the table (making sure I captured the UPC of each item) and then as I was walking back to the boardroom spoke this prompt into ChatGPT with the uploaded photos; "we need to reply to an RFP. Can you review all the items in this photo and create a pricing table for this kit that includes the item number, UPC, description, and color." The output was near perfect. When I pulled it up on screen and added the following prompt; "now, can you put it in excel so our team can cut and paste it to our template" is when the CEO said, "Dude, this is fucking crazy..."

That single demonstration saved his team roughly 6 hours per RFP. When you're responding to 40-50 RFPs per month, you're talking about 240-300 hours of manual data entry that just evaporated. But the real breakthrough wasn't the time savings - it was watching this CEO's entire worldview shift in real time.

He went from seeing AI as some abstract tech buzzword, to understanding it as a practical business tool that could solve actual problems his team faced every day. More importantly, he started connecting the dots between what we'd just demonstrated and every other process bottleneck in his operation.

"Wait," he said, leaning forward in his chair, "if it can do this with photos and create spreadsheets, what else can it handle? What about our customer service responses? What about training materials for new hires? What about analyzing our win/loss data?"

That's when I knew I was witnessing something bigger than just another software implementation. This was a CEO having his "internet moment" - that exact instant when you realize the ground has shifted underneath your entire industry and everything you thought you knew about how business gets done is about to change.

Over the next six months, his company transformed in ways that still surprises me. Their sales team's response time to RFPs dropped from days to hours. Their close rate on government contracts increased by 23%. Their newest sales reps were performing at levels that used to take years to achieve. Their entire approach to customer acquisition evolved from reactive quote responding, to proactive relationship building.

Here's what really matters: while all this was happening, their competitors were still doing things the old way. Still manually recreating pricing tables. Still sending generic outreach emails. Still treating each RFP like a completely unique challenge instead of a systematic process that could be optimized.

The competitive advantage wasn't just about efficiency - it was about fundamentally superior capability. They weren't just doing the same things faster; they were doing things their competitors literally couldn't do.

That's when I realized most business leaders were asking the wrong question about AI. They kept asking, "What can AI do?" when they should have been asking, "What becomes possible when AI eliminates our biggest bottlenecks?"

The difference between those two questions is the difference between incremental improvement and transformational change. It's the difference between using AI as a fancy calculator and using it as the foundation for an entirely new way of operating.

What You're Actually Reading

Before we go any further, I need to be completely transparent about something: this book was written in partnership with artificial intelligence. Not because I couldn't write it myself - I've been writing about sales and business strategy for over two decades - but because what kind of AI business book would this be if it weren't actually created using the technology it's teaching?

Every framework you'll encounter came from real client work. Every insight emerged from actual implementations I've personally overseen. Every recommendation has been battle-tested in the field.

The AI helped me organize these experiences into coherent systems, research supporting data, and maintain consistency across what became a surprisingly complex web of interconnected strategies.

Think of it this way: I brought the experience, the judgment calls, and the "this actually works in real businesses" perspective. The AI brought research capability, organizational structure, and the ability to synthesize years of scattered insights into something you can actually implement.

This is exactly the kind of human-AI collaboration I'll be teaching you to build in your own organization.

About the Stories You'll Read

Throughout this book, you'll encounter detailed case studies of companies that have implemented AI sales systems. Some have come directly from my client work, however, I've changed identifying details to protect proprietary information and competitive advantages that my clients have rightfully earned.

In other cases, I've created composite examples that represent common patterns I've observed across multiple implementations. This approach lets me show you more comprehensive scenarios than any single client could provide, while protecting the confidentiality every business leader expects when working with a consultant.

Every challenge described is real. Every solution has been tested in actual business environments. Every result cited reflects genuine outcomes - even when specific details have been adapted to protect the companies involved.

What's Coming Next

The AI Sales Engine framework didn't emerge overnight. It evolved from systems I'd been developing around customer acquisition and sales optimization; then got completely transformed when I started experimenting seriously with AI tools in late 2022.

What you're about to discover is a systematic approach to building sales operations that learn, adapt, and scale in ways that were literally impossible just two years ago. It's organized into three phases - Architect, Assemble, and Activate - that will take you from wherever you are now to a fully operational AI-enhanced revenue machine.

This isn't about replacing your sales team with chatbots. It's about giving your existing people capabilities that make them dramatically more effective at every stage of the sales process.

But here's what I really need you to understand: the companies implementing these systems now, while their competitors are still debating whether AI is a real business tool or just hype, are building advantages that compound every single day. The companies waiting for the technology to "mature further" will find themselves trying to catch up to competitors who are already operating with fundamentally superior capabilities.

That moment in the boardroom when the CEO said "this is fucking crazy" wasn't really about ChatGPT or RFP processing. It was about recognizing when the entire game is changing. It was about having the awareness to lean into uncertainty when the potential upside justifies the risk.

We're not just watching sales technology evolve. We're witnessing the birth of an entirely new category of business capability. The question isn't whether this transformation will happen - it's already happening. The question is whether you'll lead it, or get left behind by it.

Welcome to the AI Sales Engine.

Doug Foley CEO, FoleyMedia.ai Host, 9 Figure Exits Podcast

AI DISCLAIMER AND WAIVER

AI-Assisted Content Creation

This book was created through a collaborative partnership between human expertise and artificial intelligence technology. The frameworks, strategies, methodologies, and business insights contained herein are derived from the author's extensive professional experience in sales, marketing, and business consulting. However, the organization, research, writing, and presentation of this content was significantly enhanced through the use of AI tools and technologies.

The AI assistance included but was not limited to:

- Content organization and structure
- Research and data compilation
- Writing assistance and editing
- Case study development and presentation
- Technical explanations and clarifications

While the core knowledge and strategic insights originate from real-world business experience, the final presentation represents a human-AI collaborative effort designed to demonstrate the practical applications of the technologies discussed within the book itself.

Case Study Modifications and Confidentiality Protection

Many of the case studies, client examples, and business scenarios presented in this book have been modified, adapted, or anonymized to protect the confidentiality and proprietary information of actual clients, partners, and business relationships. These modifications may include but are not limited to:

- Changes to company names, sizes, and industries
- Alterations to specific financial figures, metrics, and performance

- Modifications to geographic locations, market segments, and
- Adjustments to timelines, implementation periods, and project scopes
- Combination of multiple client experiences into composite examples

While these case studies are based on real business situations and actual outcomes, the specific details presented may not reflect the exact circumstances of any individual client engagement. The purpose of these modifications is to illustrate proven principles and methodologies, while respecting confidentiality agreements and protecting competitive advantages.

Third-Party Case Studies and Citations

In addition to modified client examples, this book incorporates case studies, research data, and business examples from publicly available third-party sources. These external case studies have been cited and attributed where appropriate and are used to supplement and illustrate the principles and strategies discussed.

The inclusion of third-party case studies does not imply endorsement of the companies, products, or services mentioned, nor does it suggest any business relationship between the author and the organizations cited.

Disclaimers and Limitations of Warranties

NO WARRANTIES OR REPRESENTATIONS: The author makes no warranties, representations, or guarantees regarding the accuracy, completeness, or reliability of any information presented in this book, including but not limited to:

- Financial figures, performance metrics, or ROI calculations
- Specific implementation timelines or expected outcomes

- Market data, industry statistics, or competitive analyses
- Technical specifications or software capabilities
- Legal, regulatory, or compliance considerations

EDUCATIONAL PURPOSE: This book is intended for educational and informational purposes only. The strategies, methodologies, and recommendations presented are based on the author's professional experience and observations but may not be suitable for every business situation, industry, or organizational context.

NO GUARANTEE OF RESULTS: Past performance and case study results do not guarantee future outcomes. Business results depend on numerous factors including but not limited to market conditions, execution quality, organizational capabilities, competitive dynamics, and economic factors beyond the control of any methodology or framework.

PROFESSIONAL ADVICE DISCLAIMER: The content in this book does not constitute professional business advice, legal counsel, financial guidance, or technical consultation for any specific situation. Readers should consult with qualified professionals before implementing any strategies, technologies, or business changes discussed in this material.

Technology and Platform Disclaimers

The AI tools, software platforms, and technologies referenced in this book are subject to continuous updates, modifications, and changes by their respective providers. Features, capabilities, pricing, and availability mentioned may have changed since publication. Readers should verify current information directly with technology providers before making implementation decisions.

The author does not warrant the continued availability, functionality, or performance of any third-party technologies, platforms, or services mentioned in this book.

Liability Limitations

LIMITED LIABILITY: To the fullest extent permitted by law, the author, publisher, and any affiliated parties shall not be liable for any direct, indirect, incidental, consequential, or punitive damages arising from the use of this book or the implementation of any strategies, recommendations, or methodologies contained herein.

ASSUMPTION OF RISK: Readers acknowledge that business strategy implementation involves inherent risks and uncertainties. Any decisions made based on the information in this book are made at the reader's own risk and discretion.

INDEMNIFICATION: Readers agree to indemnify and hold harmless the author, publisher, and affiliated parties from any claims, damages, losses, or expenses arising from their use of this book or implementation of its contents.

Intellectual Property and Usage Rights

The frameworks, methodologies, and systems presented in this book, including the AI Sales Engine, represent proprietary intellectual property developed through professional experience and research. While readers are encouraged to implement these strategies in their own businesses, any commercial use, republication, or distribution of the specific methodologies requires express written permission.

The AI tools and technologies referenced remain the intellectual property of their respective owners and are subject to their individual terms of service and licensing agreements.

Updates and Modifications

Given the rapidly evolving nature of AI technology and business applications, some information in this book may become outdated or require modification. The author reserves the right to update, modify, or supplement this content through future editions, online resources, or other communication channels.

Readers are encouraged to verify current best practices, technology capabilities, and market conditions when implementing any strategies discussed in this book.

By reading and using this book, you acknowledge that you have read, understood, and agreed to these disclaimers and limitations. If you do not agree with these terms, please discontinue use of this material.

Last Updated: March 2026

INTRODUCTION

Why Most Companies Fail at AI-Driven Growth (and How to Do It Right)

Every company I walk into has one. Sometimes it's a Salesforce dashboard that cost $80,000 to build and hasn't been opened since the demo. Sometimes it's a conversation intelligence platform that three people use - and two of them only log in because they get reminder emails. Sometimes it's an AI chatbot that went live with great fanfare in Q1 and got quietly turned off by Q3 because it was annoying more prospects than it was converting.

I call it the AI graveyard. And if you're running a B2B sales organization right now, you've got one too.

The tools aren't the problem. Most of them actually work. The prospecting platforms find real prospects. The email automation sends real emails. The analytics dashboards display real data. But none of them talk to each other. None of them are connected to a coherent strategy. And nobody on your team can explain how Tool A is supposed to feed into Tool B, which is supposed to trigger Tool C, which is supposed to generate pipeline.

So what happens? Your team reverts to what they know. They use the CRM as a contact database and ignore the AI features. They write emails from scratch instead of using the sequences someone spent three weeks building. They eyeball their pipeline instead of trusting the predictive scoring model. And every month, you write another check for platforms that are basically expensive shelf decorations.

This isn't a technology failure. It's an architecture failure. And it's costing companies millions - not just in wasted software spend, but in the revenue they should be generating and aren't.

The companies that figure this out - the ones that stop buying tools and start building systems - see a completely different reality. Pipeline velocity that doubles and triples. Sales cycles that compress by a third. EBITDA growth that shows up in months, not years. Not because they found better AI tools, but because they connected the ones they already had into something that actually works.

That's what this book is about. Not another list of AI tools you should buy. A framework for building an AI Sales Engine that turns your existing technology, data, and team into a predictable revenue-generating system.

The $12.8 Billion Missed Opportunity

According to Grand View Research (2024), the global sales enablement platform market will reach $12.8 billion by 2030, growing at a compound annual growth rate of 16.3%. HubSpot's 2024 AI Sales Trend Report indicates that AI adoption in sales has doubled from 24% to 43% in just one year (HubSpot, 2024). Research from SaaS Capital (2025) shows that 76% of private SaaS companies are now using AI in their operations, with significant impacts on profitability and operational efficiency.

Despite this massive investment and adoption, most companies are failing to realize the promised returns. They're treating AI like a magic solution, rather than what it actually is: a powerful tool that requires precise architecture, systematic implementation, and disciplined execution.

The problem isn't the technology. The problem is the approach.

Why Traditional Sales Is Losing Ground

The sales landscape has fundamentally changed. Prospects are more informed, more skeptical, and more overwhelmed than ever before. They're conducting 73% of their buying research before ever speaking to a salesperson. They're receiving an average of 145 sales emails per week. Traditional spray-and-pray tactics don't just fail - they actively damage brands.

Meanwhile, companies that have successfully implemented AI-driven sales engines are experiencing:

- 300% increases in qualified pipeline
- 45% reductions in sales cycle time
- 67% improvements in conversion rates
- 23% increases in average deal size

The gap between AI-powered sales organizations and traditional sales teams isn't just widening - it's becoming an insurmountable competitive advantage.

Who This Book Is For

This book was written specifically for Entrepreneurs, CEOs, Chief Revenue Officers, Chief Marketing Officers, and executives at:

- Well-funded startups ready to scale beyond founder-led sales
- PE-backed B2B companies targeting accelerated growth or strategic
- Growth-stage enterprises ($10M-$100M+ revenue) seeking competitive
- Fortune 100 organizations needing systematic approaches to AI sales

If you're responsible for revenue growth, stakeholder returns, or strategic positioning in your organization, this book will provide the frameworks, tools, and methodologies to build a sustainable competitive advantage through AI-powered sales.

Why We're "Architects," Not Consultants

Before we go further, let me be clear about our positioning and approach. At FoleyMedia.ai, we don't call ourselves consultants. We're architects.

Consultants diagnose problems and make recommendations. Architects design systems, build foundations, and create structures that last. The difference is crucial when it comes to AI sales enablement.

Most companies fail at AI-driven sales because they're approaching it like a consulting project - they're looking for quick fixes, silver bullets, and tactical implementations. They're buying tools without building systems. They're implementing technology without creating architecture.

The AI Sales Engine methodology you'll learn in this book is based on the following architectural principles:

Foundation First: Before you select a single AI tool, you must build a precise understanding of your ideal customer profile, messaging framework, and revenue architecture.

Systematic Integration: Technology serves the system, not the other way around. Every tool, platform, and automation must integrate seamlessly into your overall revenue engine.

Sustainable Scale: Quick wins are worthless if they can't be replicated, improved, and scaled. Everything we build is designed for long-term competitive advantage.

Measurable Impact: Architecture requires blueprints, and blueprints require measurements. Every component of your AI Sales Engine must be trackable, optimizable, and directly tied to revenue outcomes.

What You'll Learn

This book will teach you how to architect, assemble, and activate an AI Sales Engine that delivers measurable, sustainable revenue growth. Specifically, you'll discover:

The 3-Phase AI Sales Engine Framework: A proven methodology for transforming your sales organization into a predictable revenue-generating system.

Precision ICP Development: How to move beyond generic buyer personas to create laser-focused ideal customer profiles that drive every aspect of your sales strategy.

AI Stack Selection & Integration: A systematic approach to evaluating, selecting, and integrating AI tools that actually improve results rather than adding complexity.

Bulletproof Implementation Systems: The standard operating procedures, training frameworks, and management systems required to successfully deploy AI at scale.

Performance Optimization Methodologies: How to continuously improve your AI Sales Engine performance using data-driven optimization techniques.

ROI Measurement & Board Reporting: The metrics, dashboards, and attribution models that prove the business impact of your AI investments.

The Origin Story: From Happiness of Pursuit to AI Sales Architecture

My journey to AI sales architecture wasn't traditional. I started as the host of "The Happiness of Pursuit" podcast, interviewing entrepreneurs about following their passions to create extraordinary businesses. I then wrote "Breakout Blueprint (2021)," teaching founders how to build profitable lifestyle businesses.

But as I worked with more scaling companies, I noticed a pattern: The entrepreneurs who achieved the largest exits weren't necessarily the most passionate or visionary; they were the most systematic. They built frameworks, created processes, and implemented systems that could scale beyond their personal involvement.

This insight led me to develop the Customer Loyalty Framework - a systematic approach to building sustainable, profitable customer relationships. The framework worked so well that private equity firms began asking me to implement it in their portfolio companies.

That's when I discovered the gap between strategy and execution that plagues most growing companies. Leaders understood what needed to be done, but they lacked the systems, tools, and processes to do it consistently at scale.

The rise of AI changed everything. Suddenly, the systematic approaches that had previously required large teams and significant resources, could be automated, optimized, and scaled using intelligent technology. However, most companies were implementing AI tactically rather than strategically - buying tools instead of building systems.

This led to the development of the AI Sales Engine methodology and the founding of FoleyMedia.ai. Today, we work exclusively with growth-stage companies, private equity firms, and strategic buyers who need systematic approaches to AI-powered revenue generation.

How to Use This Book

This book is designed as both a strategic guide and a practical implementation manual. Each section builds on the previous one, so I recommend reading it sequentially the first time through.

Section 1 establishes the strategic context and introduces the AI Sales Engine framework.

Sections 2-4 provide detailed implementation guidance for each phase of the methodology (Architect, Assemble, Activate).

Section 5 covers results measurement, implementation management, and future-proofing strategies.

Section 6 addresses ethical considerations and provides expert insights from industry leaders.

Throughout the book, you'll find:

- **Case Studies**: Real-world examples of successful AI Sales Engine
- **Templates & Checklists**: Practical tools you can use immediately
- **Executive Takeaways**: Boardroom-ready summaries of key concepts
- **Implementation Timelines**: Realistic roadmaps for deploying the

A Note on Results

The strategies, frameworks, and methodologies in this book work. They've been tested with dozens of companies across multiple industries, from high-growth startups to Fortune 500 enterprises. In order to do so, they require commitment, discipline, and systematic execution.

AI is not a magic solution. It's a powerful amplifier. If your sales process is broken, AI will amplify the problems. If your ideal customer profile is unclear, AI will scale the confusion. If your messaging doesn't resonate, AI will automate the irrelevance.

However, if you follow the AI Sales Engine methodology - if you architect before you assemble, assemble before you activate, and optimize continuously - you'll build a revenue-generating system that creates a sustainable competitive advantage.

The question isn't whether AI will transform sales. It already has. The question is whether you'll be among the companies that harness this transformation for competitive advantage, or among those that get left behind.

Let's begin.

SECTION 1: THE AI SALES ENGINE FRAMEWORK

CHAPTER 1

Why AI Sales Engines Are No Longer Optional

In early 2024, Palantir's CEO Alex Karp had a problem most executives would kill for: so much demand for his AI platform that his company couldn't process it all. His exact words were, "We don't know what to do with the onslaught of demand."

But here's what makes Palantir's story relevant to every B2B company reading this book - it wasn't their technology that created the onslaught. It was their sales engine.

Palantir didn't build a traditional enterprise sales team. Karp once said the only way he'd hire salespeople was if he were "hit by a bus." Instead, they built something different: intensive AI boot camps - part hackathon, part demo, part hands-on workshop - where prospects could deploy Palantir's platform against their own real-world problems in a matter of days, not months. Engineers became the salesforce. Existing customers became the case studies. The product sold itself because prospects experienced the results before they ever signed a contract.

The impact was staggering. US commercial revenue jumped 121% year over year. Total contract value surged 342%. Their sales cycle compressed from months-long enterprise negotiations into days-long boot camps that converted skeptics into advocates. By late 2025, Palantir was generating nearly $1.2 billion in quarterly revenue, and their growth trajectory was accelerating, not slowing down.

Now think about Palantir's competitors - the enterprise software companies still running the traditional playbook. Still scheduling discovery calls. Still sending follow-up sequences. Still waiting six months for a procurement committee to make a decision. They're selling the same category of product to the same category of buyer, and they're getting lapped.

That's the divergence happening across every B2B industry right now. Not just in enterprise AI - in cybersecurity, fintech, manufacturing technology, professional services, SaaS. The companies that have built systematic, AI-powered approaches to revenue generation aren't just outperforming their competitors. They're redefining the game while everyone else is still debating the rules.

The question isn't whether this shift is real. The data in this chapter will make that painfully clear. The question is which side of the divide you're on - and how much time you have left to cross it.

The Death of Traditional Sales: Market Reality Check

Traditional sales methodologies aren't just becoming less effective - they're becoming counterproductive. Here's what the data tells us:

Response Rates Are Collapsing: Cold email response rates have dropped from 18% in 2018 to less than 8.5% in 2024. Cold calling success rates have fallen by 45% over the same period.

Buying Behavior Has Fundamentally Changed: B2B buyers now complete 73% of their research before engaging with sales (reference). They consume an average of 13 pieces of content before making purchase decisions (reference). The traditional sales process has been largely eliminated by the time companies have their first conversation.

Sales Cycles Are Extending: Average B2B sales cycles have increased by 23% since 2020, while the number of stakeholders involved in purchase decisions has grown from 6.8 to 11.2 people (reference).

Competition Is Intensifying: The average B2B buyer now evaluates 3.7 vendors before making a decision; up from 2.1 vendors in 2019 (reference). Differentiation based on features or pricing has become nearly impossible in most markets.

Meanwhile, companies using AI-powered sales engines are experiencing exactly the opposite trends:

- **Higher Response Rates**: AI-personalized outreach generates 29%
- **Accelerated Sales Cycles**: Intelligent lead scoring and routing
- **Increased Deal Sizes**: AI-driven upselling and cross-selling
- **Improved Win Rates**: Predictive analytics and optimized follow-up

The divergence isn't subtle. It's a complete separation between companies that have embraced AI-driven sales and those still relying on traditional methods.

How AI Transforms Revenue Acceleration

AI doesn't just improve sales efficiency - it fundamentally transforms how revenue is generated. Here are the four primary ways AI creates sustainable competitive advantage in B2B sales:

1. Precision Targeting at Scale

Traditional sales teams cast wide nets and hope for the best. AI-powered sales engines identify the precise characteristics of buyers most likely to purchase, when they're most likely to buy, and what messaging will resonate most effectively.

The results are striking across the industry. Companies implementing AI-driven lead scoring move from broad targeting - "mid-market companies with 50-500 employees" - to precision identification of specific characteristics that predict purchase likelihood. Salesforce research shows that eighty-four percent of sales teams using AI report increased sales, with the most significant gains coming from better targeting and prioritization rather than just increased volume.

2. Intelligent Personalization

AI enables personalization that would be impossible with traditional sales methods. Instead of generic outreach templates, AI-powered systems can generate customized messaging based on company news, technology stack, funding events, personnel changes, and dozens of other dynamic factors.

Case Example: A B2B SaaS company replaced their generic email sequences with AI-generated personalized outreach. Each prospect received messaging tailored to their specific industry challenges, competitive landscape, and growth stage. Response rates increased from 6.2% to 18.7% within 90 days.

Source: HubSpot campaign analytics and customer interview validation, Q3 2024.

3. Predictive Revenue Intelligence

AI transforms sales from a reactive process to a predictive one. Instead of waiting to see which leads convert, AI engines predict conversion probability, optimal timing, pricing sensitivity, and likely objections before the first sales conversation.

Predictive analytics transforms resource allocation from gut feel to mathematical probability. Instead of pursuing every opportunity equally, AI engines predict conversion probability based on hundreds of variables - allowing sales teams to focus intensive effort on high-probability deals while maintaining lighter engagement with longer-term opportunities. The shift from reactive pipeline management to predictive resource allocation consistently produces dramatic improvements in quarterly attainment.

4. Continuous Optimization

Traditional sales improvement relies on quarterly reviews and annual planning cycles. AI-powered sales engines optimize continuously, testing messaging variations, adjusting targeting parameters, and refining processes in real-time based on performance data.

Case Example: A financial services company deployed AI-powered A/B testing across their entire sales process. The system automatically tested email subject lines, call timing, proposal formats, and follow-up sequences. Over 12 months, continuous optimization improved overall sales performance by 156%.

Source: CRM performance analytics and A/B testing results analysis, Q2 2023-Q2 2024.

The Risk of Ignoring AI in Your Sales Strategy

The risks of not implementing AI in your sales strategy extend far beyond missed opportunities. Companies that fail to adopt AI-driven sales methodologies face four critical threats:

1. Competitive Displacement

Your AI-powered competitors aren't just more efficient - they're more effective. They respond faster, personalize better, and close more deals. This isn't a temporary advantage; it's a compounding competitive advantage that grows stronger over time.

2. Talent Disadvantage

Top sales professionals want to work with cutting-edge tools and systems. Companies using traditional sales methods struggle to attract and retain high-performing sales talent.

3. Scaling Limitations

Traditional sales methods require linear scaling - more revenue requires more salespeople. AI-powered sales engines enable exponential scaling - the same team can generate significantly more revenue with better tools and systems.

4. Strategic Vulnerability

Private equity buyers, strategic acquirers, and public market investors increasingly view AI-powered sales capabilities as essential for sustainable competitive advantage. Companies without sophisticated sales technology face valuation discounts and strategic disadvantages.

The Window of Opportunity

Here's the critical insight most CEOs miss: The competitive advantage of AI-powered sales isn't permanent. It's a window of opportunity that's closing rapidly.

Right now, AI adoption in sales sits at 43%. Companies that implement sophisticated AI sales engines today have a 12-18 month advantage over competitors. But as adoption accelerates, this advantage will commoditize.

The companies building AI Sales Engines today will maintain a sustainable competitive advantages. The companies waiting for AI to become "easier" or "cheaper" will find themselves permanently disadvantaged.

Chapter 1 Executive Takeaways

For the CEO: AI-powered sales isn't optional - it's table stakes for competitive relevance. The question isn't whether to implement AI, but how quickly you can build systematic advantages.

For the Chief Revenue Officer (CRO): Traditional sales methodologies are failing because buyer behavior has fundamentally changed. AI bridges the gap between how buyers want to buy and how companies need to sell.

For the CMO: AI-powered sales requires AI-powered marketing. The companies winning with AI sales have integrated AI across their entire revenue generation process.

For the Board: AI sales capabilities directly impact valuation multiples, exit readiness, and competitive positioning. This is a strategic investment, not a tactical expense.

The age of hoping for sales results is over. The age of systematically generating them has begun.

In Chapter 2, we'll introduce the AI Sales Engine framework - the systematic methodology that transforms AI from a collection of tools into a predictable revenue-generating system.

CHAPTER 2

Introducing the AI Sales Engine

Walk into any Home Depot on a Saturday morning and you'll see the same guy. He's got a cart full of power tools - a miter saw, a brad nailer, an orbital sander, a router he watched a YouTube video about. He's spending two grand on equipment because he wants to build a deck.

What he doesn't have is a blueprint. No measurements. No materials list. No plan for how the deck connects to the house, handles drainage, or meets code. He's got $2,000 worth of tools and absolutely no idea what to do with them on Monday morning.

That's exactly how most companies approach AI sales.

They buy the prospecting platform because a competitor mentioned it at a conference. They add the email automation tool because their reps are drowning in manual outreach. They sign up for conversation intelligence because the demo was impressive. They layer on predictive analytics because a board member read an article about it.

Six months later, they've got a dozen platforms that don't talk to each other, a sales team toggling between tabs all day, and results that are somehow worse than before they started spending. The tools aren't bad. Most of them actually do what they promise. But nobody designed how they're supposed to work together, in what order, or toward what outcome.

Buying AI tools without a framework is like buying every power tool in the store without a blueprint. You'll make a lot of noise. You'll spend a lot of money. And you'll end up with something that looks nothing like what you had in mind - if you end up with anything at all.

The AI Sales Engine methodology exists to solve this problem. It's the blueprint - a structured, three-phase framework that tells you exactly what to build, in what order, and how each piece connects to the next. It turns a pile of disconnected tools into an integrated system that generates revenue predictably and scales without breaking.

The 3-Phase AI Sales Engine Model Overview

The AI Sales Engine is built on three sequential phases that must be completed in order:

PHASE 1: ARCHITECT - Design your revenue foundation before you buy a single AI tool

PHASE 2: ASSEMBLE - Integrate AI technology into a coherent, scalable system

PHASE 3: ACTIVATE - Launch, optimize, and scale your AI-powered revenue engine

Each phase contains specific methodologies, deliverables, and success metrics. More importantly, each phase builds on the previous one - you cannot successfully assemble an AI sales system without proper architecture, and you cannot successfully activate campaigns without proper assembly.

Let me break down each phase:

PHASE 1: ARCHITECT (Foundation Design)

The Architecture phase focuses on understanding exactly who you sell to, why they buy, and how they make decisions - before you implement any AI technology.

Core Components:

- Precision Ideal Customer Profile (ICP) development
- Data-driven buyer persona research
- Messaging framework creation
- Revenue architecture blueprint

- Stakeholder alignment workshops

Key Deliverable: A boardroom-grade ICP Blueprint that defines your entire go-to-market strategy

Success Metric: 100% leadership alignment on target customer definition and value proposition

Why This Matters: AI amplifies everything. If your targeting is imprecise, AI will scale imprecision. If your messaging doesn't resonate, AI will automate irrelevance. The Architecture phase ensures your foundation is solid before you build on it.

PHASE 2: ASSEMBLE (System Integration)

The Assembly phase focuses on selecting and integrating AI tools that serve your specific revenue architecture - not the other way around.

Core Components:

- AI stack evaluation and selection
- CRM and marketing technology integration
- Data pipeline architecture
- Standard Operating Procedure (SOP) development
- Team training and enablement

Key Deliverable: A fully integrated AI sales system with documented processes and trained users

Success Metric: All AI tools communicate seamlessly, data flows automatically, and team adoption exceeds 85%.

Why This Matters: Integration is where most AI initiatives fail. Companies buy powerful individual tools that don't work together, creating more complexity instead of more capability. The Assembly phase ensures your AI stack functions as a unified system.

PHASE 3: ACTIVATE (Campaign Optimization)

The Activation phase focuses on launching AI-powered campaigns and continuously optimizing performance based on data-driven insights.

Core Components:

- Campaign design and launch
- A/B testing and optimization
- Performance monitoring and analysis
- Scaling and expansion strategies
- Continuous improvement processes

Key Deliverable: Predictable, scalable revenue generation with clear ROI attribution

Success Metric: Month-over-month pipeline growth with decreasing cost per acquisition

Why This Matters: AI-powered sales is not "set it and forget it." It requires continuous optimization, testing, and refinement. The Activation phase ensures your AI Sales Engine improves over time rather than degrading.

Why a Structured, Repeatable Framework Outperforms Random AI Tools and Trial/Error

Most companies approach AI implementation like they're conducting science experiments. They try one tool, then another, then another, hoping something will work. This approach fails for four reasons:

1. No Baseline for Comparison

Without a clear architecture and defined success metrics, you can't tell whether an AI tool is improving performance or making it worse. Companies end up keeping ineffective tools and discarding effective ones because they lack systematic measurement.

2. Integration Chaos

Random tool adoption creates integration nightmares. Each new tool requires custom connections, data mapping, and process changes. Instead of building capabilities, you're constantly fighting technical debt.

3. Team Resistance and Confusion

When sales teams are constantly learning new tools without understanding how they fit together, they resist adoption. The AI Sales Engine framework provides context and continuity that enables sustainable adoption.

4. Opportunity Cost

Trial-and-error AI implementation consumes enormous amounts of time, money, and organizational energy. The AI Sales Engine framework eliminates waste by providing a proven path to results.

Contrast this with the systematic approach:

Clear Architecture: Every AI tool serves a specific purpose within your overall revenue strategy.

Integrated Systems: All tools work together to amplify your sales engine rather than competing for attention.

Measured Results: Every component is tracked, tested, and optimized based on clear performance metrics.

Scalable Processes: Success can be replicated, improved, and expanded systematically

Case for Simplicity: Why Clarity in Execution Is Everything

The most successful AI Sales Engine implementations aren't the most complex - they're the most clear. Here's why simplicity drives better results:

The Complexity Trap

AI vendors love to showcase feature-rich platforms with dozens of capabilities. But complexity without clarity creates paralysis. Sales teams end up using 20% of an AI platform's capabilities instead of mastering the core functions that drive results.

Better Approach: Select AI tools that excel at specific functions within your sales process rather than platforms that attempt to do everything adequately.

The Integration Imperative

Simple, well-integrated systems outperform complex, disconnected ones every time. A basic AI tool that seamlessly integrates with your CRM is more valuable than a sophisticated platform that requires manual data entry.

Better Approach: Prioritize integration capabilities over feature lists when evaluating AI tools.

The Adoption Reality

Your sales team's adoption rate determines your AI initiative's success rate. Complex systems require extensive training, ongoing support, and constant troubleshooting. Simple systems enable immediate value creation.

Better Approach: Design your AI Sales Engine for user simplicity, even if it requires more sophisticated backend integration.

The Power of Simplicity

This plays out every day in sales technology decisions. Companies choose platforms with forty-seven features over platforms focused on three core functions because more features seem more "comprehensive." Six months later, the sales team is using only a handful of those features, data quality has degraded due to complex input requirements, and performance has actually declined. Gartner's 2024 research found that seventy percent of sellers reported being overwhelmed by the number of technologies required to do their work. Adding complexity to that overwhelm doesn't create sophistication - it creates rejection.

The most powerful AI Sales Engines are often the simplest to use. A system the team masters in two days outperforms a system that takes two weeks to learn, because adoption beats capability every time.

The Science of Predictable Revenue Growth

The AI Sales Engine framework transforms revenue generation from an art to a science. Here's how systematic AI implementation creates predictable growth:

1. Data-Driven Decision Making

Traditional sales relies on intuition, experience, and gut feelings. AI-powered sales relies on data, patterns, and statistical analysis. This shift from subjective to objective decision-making dramatically improves consistency and predictability.

Example: Instead of guessing which prospects to prioritize, AI engines analyze hundreds of data points to predict conversion probability with 89% accuracy.

2. Systematic Optimization

Traditional sales improvement happens through quarterly reviews and annual planning. AI-powered sales improvement happens continuously through automated testing and optimization.

Example: AI systems can test 15 different email subject lines simultaneously, automatically route traffic to the highest-performing variant, and generate new variations based on performance patterns.

3. Scalable Personalization

Traditional sales personalization is limited by human capacity - you can only customize so many messages, calls, and presentations. AI-powered personalization scales without limits.

Example: An AI engine can generate personalized emails for 10,000 prospects based on their specific company news, technology stack, and behavioral patterns - all delivered simultaneously.

4. Predictive Intelligence

Traditional sales is reactive - you respond to what prospects do. AI-powered sales is predictive - you anticipate what prospects will do and prepare accordingly.

Example: AI engines can predict which customers are likely to churn, which prospects are ready to buy, and which deals are most likely to close within the quarter.

The Three Pillars of AI Sales Engine Success

Every successful AI Sales Engine implementation is built on three foundational pillars:

Pillar 1: Strategic Alignment

Your AI sales strategy must align with your overall business strategy, market positioning, and growth objectives. AI is not a separate initiative - it's an integrated approach to achieving your strategic goals.

Key Questions:

- How does AI-powered sales support our competitive differentiation?
- What specific business outcomes will we measure?
- How does our AI sales strategy align with our market expansion

Pillar 2: Systematic Implementation

Your AI sales deployment must follow a systematic methodology that ensures proper sequencing, integration, and optimization. Random tool adoption creates chaos; systematic implementation creates competitive advantage.

Key Requirements:

- Phase-based deployment with clear deliverables
- Integration architecture designed before tool selection
- Change management processes for team adoption

Pillar 3: Continuous Optimization

Your AI sales engine must be designed for continuous improvement based on performance data and market changes. Static AI implementations become obsolete; dynamic AI engines become more powerful over time.

Key Capabilities:

- Real-time performance monitoring
- Automated A/B testing
- Regular strategy reviews and adjustments

Common Misconceptions About AI Sales Engines

Before we dive deeper into implementation, let's address five common misconceptions that cause AI sales initiatives to fail:

Misconception 1: "AI Will Replace Our Sales Team"

Reality: AI amplifies human capability; it doesn't replace human judgment. The most successful AI implementations enhance sales team performance rather than eliminating sales roles.

Misconception 2: "More AI Tools Equal Better Results"

Reality: Integration and optimization matter more than tool quantity. Three AI tools that work together seamlessly will outperform fifteen tools that don't integrate.

Misconception 3: "AI Implementation Is a Technology Project"

Reality: AI implementation is a business transformation project that requires strategy, process, and culture changes in addition to technology deployment.

Misconception 4: "AI Results Are Immediate"

Reality: AI Sales Engines require time to learn, optimize, and demonstrate full capability. Expect 90-120 days to see significant results and 6-12 months to achieve full ROI.

Misconception 5: "AI Works the Same for Every Company"

Reality: AI implementation must be customized to your specific market, customer base, sales process, and business model. Generic AI deployments generate generic results.

The ROI Framework: Measuring AI Sales Engine Success

The AI Sales Engine methodology includes a comprehensive ROI framework that measures success across five key dimensions:

1. Pipeline Quality Metrics

- Lead-to-opportunity conversion rate
- Opportunity-to-close conversion rate
- Average deal size progression
- Sales cycle velocity

2. Sales Efficiency Metrics

- Revenue per sales rep

- Cost per acquisition
- Sales activity productivity
- Time to productivity for new hires

3. Revenue Predictability Metrics

- Forecast accuracy
- Pipeline coverage ratios
- Revenue growth consistency
- Seasonal variance reduction

4. Customer Experience Metrics

- Response time improvement
- Personalization effectiveness
- Customer satisfaction scores
- Referral and expansion rates

5. Strategic Impact Metrics

- Market share growth
- Competitive win rates
- Customer lifetime value
- Strategic account penetration

Implementation Timeline: What to Expect

Here's a realistic timeline for AI Sales Engine implementation:

Months 1-2: Architecture Phase

- Stakeholder alignment workshops
- ICP research and development
- Messaging framework creation
- Blueprint documentation

Months 3-4: Assembly Phase

- AI tool evaluation and selection
- Integration architecture design
- System configuration and testing
- Team training and enablement

Months 5-6: Activation Phase

- Campaign design and launch
- Initial optimization and testing
- Performance monitoring setup
- Results analysis and refinement

Months 7-12: Optimization Phase

- Continuous improvement processes
- Scaling and expansion
- Advanced feature implementation
- Strategic planning for next phase

Chapter 2 Executive Takeaways

For the CEO: The AI Sales Engine framework transforms AI from a technology expense into a strategic investment. Systematic implementation generates predictable ROI and sustainable competitive advantages.

For the CRO: Random AI tool adoption creates chaos; systematic AI implementation creates capability. The three-phase framework ensures your AI investment delivers measurable results.

For the CMO: AI-powered sales requires AI-integrated marketing. The Architecture phase ensures sales and marketing alignment from the beginning of your AI transformation.

For the CTO: AI sales integration requires careful technical architecture. The Assembly phase provides a systematic approach to integration that avoids common technical pitfalls.

What's Next

In the following chapters, we'll dive deep into each phase of the AI Sales Engine methodology. You'll learn exactly how to architect your revenue foundation, assemble your AI technology stack, and activate campaigns that generate predictable growth.

But first, we need to start with the foundation: defining your revenue architecture with precision that makes AI implementation possible.

Phase 1 begins with understanding exactly who you serve and why they buy. Without this foundation, even the most sophisticated AI tools will fail to deliver results.

Let's begin building your AI Sales Engine.

PHASE 1: ARCHITECT

49

CHAPTER 3

Defining Your Revenue Architecture

Picture a boardroom at a $40 million B2B software company that just missed its growth target by double digits. Screens on the wall showing pipeline dashboards. A whiteboard from last quarter's planning session that nobody erased. Five executives around the table - CEO, CRO, CMO, Head of Customer Success, and the VP of RevOps - all staring at the same numbers and seeing completely different problems.

I ask one question: "If you had to put a million dollars behind one ideal customer profile next quarter - one bet, one motion - who is it, and why?"

The CEO goes first: "Enterprise accounts. Big logos move valuation. That's the story we need for the board."

The CRO pushes back immediately: "Mid-market ops leaders. Shorter cycles, cleaner deals, and we can actually close them this quarter."

The CMO has a different angle: "Technical champions who drive product-led adoption. That's where our inbound momentum is."

CS and RevOps cut in: "Forget net-new. The fastest revenue is sitting in our existing install base. Expansion deals close in half the time at twice the margin."

Then silence. Not the comfortable kind. The kind where everyone realizes they've been running four different plays with one budget.

That's the moment I call the audible. "We're not leaving this building until we agree on one ICP, one buying-committee map, and one motion. Same team, same week - we're doing a Stakeholder Alignment Workshop."

What happens next is intense and fast. Whiteboards fill up with every deal from the last two years - won, lost, expanded, churned. We tear apart each one. What was the trigger that made them buy? Who actually signed? How long did it take? Where did deals stall and why? We map jobs-to-be-done in the buyer's own words, not our marketing language. Anti-personas surface - the prospects who look perfect on paper but waste six months of your pipeline before going dark. Procurement realities and security requirements get named out loud instead of discovered at the finish line.

Forty-eight hours later, the team signs a one-page ICP Blueprint. Not a 40-page persona deck that nobody reads - one page. RevOps loads the qualification fields into the CRM. Marketing and sales align on a single outbound sequence. Leadership sees one dashboard instead of four. Within weeks, the early signals start showing: better meetings, fewer stalls, pipeline that actually moves.

Until your team agrees on who you serve, every dollar you spend on AI tools is split four ways - and so is your chance of winning. Bad ICP equals bad AI. Models don't create clarity. They scale whatever context you feed them. Feed them confusion, and you'll get confusion at scale.

This chapter is about making sure that never happens. We're going to Architect your revenue system from the ground up - run the Stakeholder Alignment Workshop, walk through precision persona discovery, and build the one-page ICP Blueprint that drives everything in the rest of this book. Every case study that worked in these pages started right here.

Why Precision Persona Discovery Changes Everything

Here's something most CEOs don't realize until they're six figures deep into AI tools: the technology doesn't create clarity. It amplifies whatever you feed it.

Feed a generative AI model a vague ICP like "mid-market CFOs who care about efficiency," and you'll get vague outputs at scale — thousands of generic emails that all sound the same, targeting lists that are a mile wide and an inch deep, and sales reps who still can't tell the difference between a prospect who's ready to buy and one who will waste three months of their time.

Feed that same model a precision ICP — one built from real customer data, validated through actual conversations, and aligned across your entire leadership team — and something different happens. The AI starts finding patterns you didn't know existed. It identifies companies showing buying signals you hadn't thought to look for. It generates personalized messaging that makes prospects think your rep spent an hour researching them, when it actually took six minutes.

The difference isn't the AI. It's the inputs.

Traditional buyer personas use maybe five to ten characteristics to define a target. Company size, industry, geography, revenue range, maybe technology usage. That's the equivalent of describing a house buyer as "someone who lives in the Northeast." Technically accurate. Practically useless.

AI-powered targeting can analyze two hundred or more characteristics simultaneously — specific technology stack components, recent funding activity, personnel changes in key departments, digital engagement patterns, competitive intelligence signals, content consumption behaviors, hiring velocity, even the tenure of the decision-maker in their current role. But it can only analyze these characteristics if someone defines them first. That someone is you, your leadership team, and the structured process we're about to walk through.

According to Salesforce's State of Sales Report, personalized sales experiences generate twenty-seven percent higher conversion rates than generic approaches. But here's the part most people miss: effective personalization requires fifteen times more data points than traditional sales approaches typically utilize. You can't personalize at scale with a five-bullet ICP. You need precision. And precision starts with the workshop.

The Stakeholder Alignment Workshop: What Actually Happens in the Room

I've run these workshops with companies from eight million to over a hundred million in revenue, across SaaS, professional services, manufacturing, and financial services. The dynamics are always the same, and the breakthrough always comes from the same place.

Here's what happens when you put your CEO, CRO, CMO, Head of Customer Success, and VP of RevOps in a room for a full day with no phones and one mandate: agree on who you're going after, or don't leave.

The morning starts with data, not opinions. Before anyone shares their gut feeling about who the ideal customer is, we put the numbers on the table. Every deal from the last eighteen to twenty-four months — won, lost, expanded, churned. We're looking at revenue by segment, profitability by customer type, sales cycle length, customer acquisition cost, retention rates, and expansion revenue. The data tells a story that nobody in the room has heard before, because nobody has looked at all of it together.

This is where the first surprise happens. Almost every company I've worked with discovers that their most profitable customers aren't who they think they are. The CEO might be chasing enterprise logos because they look good on a slide deck, but the data shows that mid-market companies in a specific sub-vertical close three times faster, expand at twice the rate, and churn at half the frequency. The CRO might be pushing for velocity deals, but the numbers reveal that deals involving a specific buying committee configuration — say, a CRO plus a VP of Sales

Ops plus an IT stakeholder — close at forty-two percent versus eighteen percent for single-threaded deals.

The data doesn't care about anyone's opinion. It tells you who actually buys, who stays, and who makes you money.

After the data presentation, each stakeholder gets fifteen minutes to present their version of the ideal customer. Not a polished deck — just their honest answer to the question: "If we could only sell to one type of company for the next twelve months, who would it be and why?" The differences surface fast. Marketing sees the world through inbound engagement — who's downloading content and attending webinars. Sales sees it through deal velocity — who's closing quickly with less friction. Customer Success sees it through retention and expansion — who stays, grows, and becomes a reference. Finance sees it through margins — who's profitable after you account for implementation, support, and the discount you gave to win the deal.

These perspectives aren't wrong. They're incomplete. The workshop forces the team to see the full picture.

By late morning, patterns start emerging from the customer analysis. We map the top twenty customers against every available characteristic — not just firmographics, but technographics, buying triggers, organizational structure, decision-maker profiles, and the specific business conditions that were present when they decided to buy. We look for statistical clusters. Are your best customers concentrated in a particular revenue range? Do they tend to have a specific technology in their stack? Is there a common trigger event — new funding, new leadership, a missed quarter — that preceded the purchase?

This is also where anti-personas surface, which is just as important. These are the prospects who look perfect on paper — right size, right industry, right budget — but waste enormous amounts of your sales team's time. Every company has them. Maybe it's companies in a specific sub-sector that have long procurement cycles and never close. Maybe it's organizations above a certain size where deals get stuck in committee for nine months. Naming the anti-personas is liberating. It gives your sales team permission to disqualify early and focus on prospects that actually convert.

By afternoon, the team is building the ICP hypothesis. This isn't a brainstorming session — it's a synthesis of everything the data revealed, filtered through the collective experience of the people in the room. The hypothesis includes specific firmographic criteria (revenue range, employee count, industry vertical, geography), technographic indicators (what's in their stack, what's missing, what they recently changed), buying committee composition (which titles need to be involved, who has veto power, who's the internal champion), trigger events (what business conditions create the buying window), and the specific pain points that your solution addresses for this particular profile.

We test the hypothesis against historical data on the spot. If the ICP criteria would have predicted your best twenty customers, you're on the right track. If it misses half of them, something is off and we adjust.

The last two hours are about messaging and implementation. Once the team agrees on who the ideal customer is, we build the messaging framework — not a list of features and benefits, but a map of what matters to each stakeholder at each stage of their buying process. What does the CRO at a fifty-million-dollar SaaS company care about in the first email? What about the VP of Sales Ops during the evaluation? What does the CFO need to see before signing? Each role gets specific messaging tied to their priorities, concerns, and the language they actually use.

The workshop ends with three deliverables. First, a one-page ICP Blueprint that every person in the room has signed off on — literally signed. Not a forty-page persona deck. One page with the specific criteria, scoring weights, and qualification questions that define your ideal customer. Second, a messaging matrix that maps stakeholder roles to buying stages to specific value propositions and proof points. Third, a ninety-day validation plan with clear metrics — response rates, meeting conversion, pipeline velocity — that will tell you within one quarter whether the ICP is accurate.

Why This Process Drives Hundreds of Millions in Revenue

I want to be direct about something. This isn't a theoretical exercise. The Stakeholder Alignment Workshop and the ICP Blueprint that comes out of it have been the starting point for every successful AI Sales Engine implementation we've built. The companies that have gone through this process have collectively generated hundreds of millions in pipeline and contributed to nine-figure valuation gains. And the ones that skipped it — the ones who said "we already know our ICP" and jumped straight to buying tools — are the ones who spent six figures on AI platforms that amplified the wrong message to the wrong audience.

The reason is simple. When your entire leadership team is aligned on one ICP, everything downstream works. Marketing generates leads that sales actually wants. Sales pursues prospects that actually close. Customer Success onboards customers who actually retain and expand. The CRM reflects reality instead of aspiration. And when you layer AI on top of that alignment, you get compound returns — the AI finds more of exactly the right prospects, personalizes messaging based on real patterns from real wins, and your team focuses its energy on a smaller number of higher-probability opportunities.

When the team isn't aligned, AI makes things worse. Marketing optimizes for engagement with the wrong audience. Sales gets flooded with leads that don't match what they can actually close. The AI learns from bad data and gets better at targeting bad-fit prospects. You end up spending more money to generate more pipeline that closes at a lower rate. I've seen it happen. It's expensive and demoralizing.

The workshop prevents this. It's eight hours of discomfort that saves twelve months of dysfunction.

Step-by-Step Persona Discovery: How to Gather the Intelligence You Need

The workshop creates alignment. Now you need the intelligence to validate and enrich that alignment with real-world data. This is the persona discovery process — a structured methodology that turns your ICP hypothesis into a battle-tested blueprint.

Starting with your own data. The first step is a deep analysis of your existing customer base. Pull eighteen to twenty-four months of CRM data and look at it from every angle — revenue contribution, profitability after accounting for support and implementation costs, sales cycle length, expansion patterns, and churn rates. What you're looking for are the characteristics that separate your best customers from your average ones and your worst ones. The tools for this analysis are straightforward: your CRM's reporting capabilities, a business intelligence tool like Tableau or Looker if you have one, and your customer success platform's retention and engagement data.

The patterns that emerge are often surprising. One company we worked with discovered that their fastest-closing, highest-retaining customers all had one thing in common: they had a dedicated revenue operations function. Not a specific technology, not a specific revenue range — a RevOps team. That single insight reshaped their entire ICP and tripled their outreach effectiveness within a quarter.

Understanding the competitive landscape. The second step is market and competitive research. You need to understand where your ICP fits within the total addressable market, how competitors are positioning against the same prospects, and what technology adoption patterns look like in your target segments. Industry research from Gartner, Forrester, or IDC provides market context. Competitive intelligence platforms help you track competitor positioning and messaging. The goal isn't to create a hundred-page market report — it's to answer three questions: How big is the opportunity within our ICP? Who else is going after them? And what's our differentiation in their eyes, not ours?

Talking to real customers. The third step is customer interviews, and this is where most companies either skip the work or do it badly. You need at least twenty structured interviews across your customer segments — best customers, good customers, churned customers, and prospects who evaluated but didn't buy. The questions aren't about your product. They're about the customer's world. What was happening in their business when they started looking for a solution? Who else was involved in the decision? What criteria did they use to evaluate options? What almost made them choose someone else? What would happen if they couldn't use your solution anymore?

The interviews surface insights that no amount of data analysis can reveal. You hear the language buyers actually use to describe their problems — not the language your marketing team invented. You learn about buying triggers that don't show up in CRM data — a conversation at a conference, a board mandate, a competitor win that rattled the executive team. You discover that the "champion" in the deal wasn't the person who signed the contract, but the VP of Operations who quietly lobbied for six weeks before anyone in your organization knew about it.

Validating with real outreach. The fourth step is testing the ICP against reality. Take your refined ICP criteria, build a small prospect list of one hundred to two hundred companies that match, and run a targeted outreach campaign. Track everything — response rates, meeting acceptance, qualification rates, sales cycle velocity. If the ICP is accurate, you'll see meaningfully better results than your historical benchmarks within thirty to sixty days. If it's not, the data will tell you where to adjust.

Building the One-Page ICP Blueprint

The output of all this work — the workshop alignment, the data analysis, the competitive research, the customer interviews, the validation campaigns — is a single document that becomes the foundation for everything your AI Sales Engine does.

The ICP Blueprint isn't a persona deck. It's not a slide with a stock photo of "Marketing Mary" and a list of generic pain points. It's a precision targeting document that captures exactly who you're going after and why.

The blueprint covers four areas. The first is company characteristics — the firmographic and technographic profile that defines your ideal account. This includes specific revenue ranges, employee counts, industry classifications, geographic parameters, growth stage indicators, and the technology stack patterns that predict fit. These aren't generic ranges. They're the specific criteria that emerged from your customer analysis and were validated through outreach testing.

The second area is decision-maker profiles. For each member of the typical buying committee, the blueprint captures their title and level, departmental affiliation, key responsibilities and goals, the performance metrics they're measured on, their communication preferences, and the specific concerns they'll raise during evaluation. This is what makes AI personalization possible — when you know that the VP of Sales Ops at a Series B SaaS company cares most about rep ramp time and forecast accuracy, your AI can generate messaging that speaks directly to those priorities.

The third area is the buying process itself. How does this type of company actually make a purchasing decision? What triggers the initial search? How do they evaluate options? Who has veto power? What does their procurement process look like? How long does it typically take from first conversation to signed contract? Understanding this process prevents your sales team from being surprised by committee reviews, legal holds, or budget cycles that delay deals.

The fourth area is the messaging framework. For each stakeholder, at each stage of the buying process, the blueprint maps the primary value proposition, supporting proof points, likely objections, and the specific language that resonates. This is the Messaging Matrix — not a list of features, but a map of conversations. When your SDR reaches out to a CRO, they know exactly what to say. When your AE presents to the CFO, they know exactly what business case to build. When the AI generates a personalized email, it pulls from messaging that's been validated against real customer feedback, not guessed at by a marketing team.

The blueprint should fit on one page — the summary version that gets taped to the wall above every rep's desk. Behind it sits the detailed documentation: the data analysis, the interview transcripts, the competitive positioning, the validation results. But the one-page version is what drives daily decisions. Is this prospect in our ICP? Check the page. What should this email say? Check the page. Should we pursue this deal or disqualify it? Check the page.

Chapter 3 Executive Takeaways

For the CEO: Revenue architecture is the foundation of AI sales success. Without precise ICP definition and stakeholder alignment, AI tools will amplify existing problems rather than solving them. The Stakeholder Alignment Workshop is eight hours that prevents twelve months of misalignment. Every company that has driven outsized returns with an AI Sales Engine started here.

For the CRO: The stakeholder alignment workshop is essential for AI implementation success. Misaligned teams cannot execute integrated AI strategies effectively. The process surfaces the data that settles internal debates about targeting — not opinions, not gut feel, but actual performance data from your own deals.

For the CMO: Precision persona development enables AI-powered personalization at scale. Generic personas limit AI effectiveness and waste technology investments. The ICP Blueprint gives marketing and sales a shared language and shared definition of success.

For the VP of Sales: The ICP Blueprint becomes the foundation for all sales activities. Clear targeting and messaging improve both individual rep performance and AI tool effectiveness. Your reps will know exactly who to pursue, what to say, and when to disqualify — which means more time on the right opportunities and less time chasing deals that were never going to close.

Next Steps: From Architecture to Intelligence

With your ICP Blueprint complete and stakeholder alignment achieved, you're ready to move to the next component of the Architecture phase: building data-driven audience intelligence.

In Chapter 4, we'll explore how to leverage firmographics, technographics, and intent signals to create AI-powered audience intelligence that transforms your ICP Blueprint into actionable targeting strategies.

Architecture is your foundation. Intelligence is your advantage.

References:

Salesforce. (2024). *State of sales report: 9th edition.* Salesforce Research. <https://www.salesforce.com/resources/research-reports/state-of-sales/>

CHAPTER 4

Data-Driven Audience Intelligence

Most sales teams operate on one layer of intelligence - firmographics. Company size, industry, location, maybe revenue range. That's like trying to sell a house by knowing only the buyer's zip code. You'll find people who live in the right area, but you'll have no idea who's actually looking to move, what they can afford, or when they need to be in a new place.

When you ask most companies "who are we targeting?", the answer sounds like a filter on a database: "Mid-market SaaS companies in North America with 200 to 2,000 employees." That answer creates spray-and-pray outreach - lots of activity, not many decisions. This chapter is about the two missing layers that turn targeting from a guessing game into an orchestrated revenue motion.

Here's how the three layers work:

The first layer is firmographics - the company layer. Size, industry, geography, tech stack baseline. It's necessary, but it's not sufficient. It tells you where to look. It doesn't tell you who to talk to or when.

The second layer is the person and relationship graph. Function, seniority, professional affiliations - McKinsey alumni, HBS graduates - past employers, networks, tenure signals. This is the layer that tells you not just which company to target, but which human being inside that company is the right conversation.

The third layer is trigger and intent signals - the timing layer. Funding rounds, leadership changes, aggressive hiring patterns, new system rollouts, event attendance, regulatory deadlines, content engagement. This is the layer that tells you when that person is actually ready to have the conversation.

When you stack person and timing on top of company, you stop guessing and start orchestrating.

Let me show you what this looks like in practice.

We had a client running a national executive dinner series - intimate events every six weeks designed to get decision-makers from billion-dollar-plus companies into a room together. Not a webinar. Not a conference. A curated dinner where every seat mattered. The objective was simple: fill each dinner with exact buyers, then convert the room into meetings and pipeline.

Firmographics alone would have given us a list of big companies. That list would have been useless - you can't fill a 20-person dinner with "companies." You need specific humans who will actually show up and who are actually in a position to buy.

So we stacked the layers.

For the person layer, we filtered by affiliation - McKinsey alumni and Harvard Business School graduates, because those networks carry weight and create instant peer credibility in the room. We filtered by role and tenure - new in seat less than twelve months, or recently promoted decision-makers, because people in new roles are actively looking for solutions and have the mandate to make changes.

For the timing layer, we filtered by capital signals - funding within the last twelve to eighteen months above a certain threshold, because companies with fresh capital have budget to solve problems. We looked for change indicators - new technology deployments, job postings in specific functions, regional expansion. And we layered in event adjacency - prior attendance at peer-level events, existing relationships with known champions in our network.

Three layers stacked together gave us a verified list of decision-maker emails and direct dials for people who matched on company, person, AND timing.

Then we orchestrated the outreach. We tailored invite copy by affiliation - "For McKinsey alumni" - and by role - "CFO track," "Ops track." We ran a cadence: formal invite, reminder, personal nudge from the host, LinkedIn direct message. Each dinner featured a thirty-minute talk built around the Architect, Assemble, Activate framework, a fifteen-minute live teardown, fifteen minutes of Q&A, and we booked follow-up meetings before anyone left the room. Within forty-eight hours, every attendee received a recap and a mutual plan invite. Every meeting tracked to opportunities.

Three events over two quarters. Sixty perfectly targeted attendees, all from billion-dollar-plus companies. Eight figures in pipeline sourced. Seven figures closed to date.

Before this approach, the client was measuring "reach" - how many people saw their content. After, they were measuring revenue - how many buyers were in the room and how many converted. The difference wasn't a better email. It was better intelligence.

In the pages that follow, we'll show you how to build all three layers - source the right signals, keep them compliant and current, and plug them into your go-to-market motion so that leadership sees activity roll up to pipeline and pipeline roll up to revenue.

The Three Layers in Depth: How Intelligence Becomes Revenue

The dinner series example illustrates how the three layers work together. But let me break down each layer individually so you can see how to build them, what tools power them, and what they actually reveal about your prospects that single-layer targeting misses entirely.

The company layer — firmographics. This is what most companies stop at, and it's the reason most outbound programs underperform. Firmographic data tells you the basics: company size, industry classification, geographic location, revenue range, growth stage, funding status. It's the equivalent of knowing someone's address and income bracket. Useful for filtering out the obviously wrong prospects, but nowhere near sufficient for knowing who's actually ready to buy.

Where firmographics get interesting — and where most companies leave value on the table — is when you go beyond the standard database fields and start looking at what I call advanced firmographic signals. Recent leadership changes signal organizational disruption and new priorities. Restructuring or layoff patterns indicate budget pressure or strategic pivots. Merger and acquisition activity — in either direction — creates windows where decision-makers are actively reevaluating their vendor stack. Financial performance indicators like consecutive quarters of growth or a sudden revenue dip tell you about budget availability and urgency. Even partnership announcements and vendor relationship changes can signal openness to new solutions.

The reason these advanced signals matter is that AI platforms can analyze thousands of them simultaneously, identifying patterns and correlations that no human analyst would catch. A traditional sales team might notice that a company just raised a Series B. An AI-powered intelligence system notices that a company raised a Series B, hired a new CRO two months later, posted five sales roles in the last three weeks, recently added Salesforce to their tech stack, and is headquartered in a state where your existing customers cluster — and it recognizes that this specific combination of signals predicted a closed deal sixty-eight percent of the time in your historical data.

That's the difference between a filter and intelligence.

The person and relationship layer — technographics and people data. The second layer answers a question that firmographics can't: who inside this company is the right conversation, and what do we know about their world?

Technographic intelligence reveals the technology stack a company uses — and that stack tells a story about their sophistication, their budget, their priorities, and their pain points. A company running Salesforce plus Outreach plus ZoomInfo plus Gong is operating at a completely different level than a company using spreadsheets and Gmail. The first company has budget, values sales technology, and is likely hitting the ceiling of what their current tools can do. The second company isn't ready for what you're selling — or if they are, they need a completely different conversation.

But technographics go deeper than just "what tools do they use." The real value is in the patterns. What technology did they recently add? If they just deployed a new CRM, they're in build mode — open to complementary tools, actively thinking about their stack, and spending money. What technology did they recently remove? If they just sunset a competitor's product, that's a massive opportunity window. What's missing from their stack? If they have a CRM and email automation but no personalization layer, no conversation intelligence, no data enrichment — those gaps are exactly where your solution fits.

The people layer maps specific human beings to these companies. Not "the VP of Sales" as an abstract concept — Sarah Johnson, VP of Sales, joined the company four months ago from a competitor where she used your product, has a Harvard Business School MBA, previously worked at McKinsey, and just posted about sales process challenges on LinkedIn last week. That level of specificity is what turns a cold email into a warm conversation. It's the difference between "Hi, I noticed you work in sales" and "Sarah, you scaled the team at CompetitorCo from eight to twenty-five reps — are you building the same motion here, and if so, how are you handling personalization at that volume?"

The dinner series succeeded precisely because we stacked these layers. We didn't just find "executives at big companies." We found McKinsey alumni and Harvard Business School graduates in specific operating roles — people whose professional networks and analytical frameworks made them predisposed to evaluate solutions rigorously and move quickly when they found the right one. We filtered by tenure — new in seat less than twelve months — because people in new roles are actively looking for ways to make an impact.

The timing layer — intent and trigger signals. This is the layer that turns a good prospect into an urgent one. You can have the perfect company with the perfect decision-maker, but if the timing is wrong, your outreach lands in the "interesting but not right now" bucket. Intent signals tell you when "not right now" becomes "we need to solve this yesterday."

The highest-value timing signals fall into several categories. Capital signals — recent funding, acquisition, or IPO preparation — indicate budget availability and growth mandate. Leadership changes — a new CRO, a new CEO, a new VP of Sales — indicate someone with a mandate to make changes and prove their impact in the first ninety days. Hiring signals — specifically what roles they're posting and how aggressively — reveal growth trajectory and organizational priorities. If a company just posted six sales roles in three weeks, they're scaling fast and about to hit every scaling problem your solution addresses.

Technology change signals are particularly powerful. When a company removes a competitor's tool from their stack, you have a narrow window where they're actively looking for a replacement. When they add a new CRM, they're in build mode and receptive to complementary tools. When they adopt a new analytics platform, it signals a shift toward data-driven decision-making — which is exactly the mindset that makes them receptive to AI-powered sales tools.

Content and engagement signals — webinar attendance, whitepaper downloads, competitor comparison searches, pricing page visits — reveal active research behavior. A prospect who visited your competitor's pricing page three times last week, downloaded a comparison guide, and attended a webinar on "choosing the right sales intelligence platform" isn't casually browsing. They're in an active evaluation cycle, and every day you don't reach out is a day closer to them choosing someone else.

In the dinner series, the timing layer was the difference between filling a room and filling it with buyers. We filtered by capital signals — funding within the last twelve to eighteen months above a specific threshold — because companies with fresh capital have both the budget and the mandate to invest. We layered in change indicators: new technology deployments, job postings in specific functions, regional expansion. And we added event adjacency — prior attendance at peer-level events — because people who attend executive dinners tend to attend more executive dinners. The timing layer didn't just tell us who was qualified. It told us who was qualified right now.

From Intelligence to Action: Making the Layers Work Together

Understanding the three layers is one thing. Operationalizing them is where most companies struggle. Let me walk you through how to actually build an intelligence system that feeds your go-to-market motion.

Choosing your intelligence platforms. The platform landscape breaks into three categories. Comprehensive data platforms like ZoomInfo provide broad firmographic, technographic, and contact data with high accuracy and enterprise-grade support — but at enterprise-grade prices, typically twenty-five to fifty thousand dollars per year on contract. Intent data specialists like Bombora and 6sense monitor content consumption and research behavior across thousands of websites, identifying which accounts are actively researching topics relevant to your solution — they're powerful for timing signals but need to be layered on top of a firmographic foundation. And enrichment orchestration platforms like Clay let you stack multiple data providers in sequence, building customized intelligence workflows that combine cheap and expensive sources for maximum coverage at minimum cost.

For most companies in the ten to one hundred million revenue range, the right approach is a hybrid. Use an enrichment orchestration platform as your primary engine — it gives you flexibility, cost control, and the ability to stack all three intelligence layers into a single workflow. Supplement with a comprehensive data provider for deep enrichment on your highest-priority accounts. And add an intent data provider if your ICP is well-defined enough that topic-based intent monitoring would meaningfully improve your timing.

The platform decision matters, but it's not the first decision. The first decision is whether your ICP Blueprint from Chapter 3 is specific enough to feed these platforms meaningful criteria. If your ICP is still "mid-market SaaS companies," no intelligence platform in the world will save you. If your ICP specifies revenue range, technology stack requirements, buying committee composition, trigger events, and the forty-seven characteristics that predict purchase likelihood — then these platforms become force multipliers.

Building the scoring model. Raw intelligence data is noise until you score it. Scoring is how you answer the most important question in sales: "Of the thousand companies that match our ICP, which twenty-five should we contact this week?"

An effective scoring model combines signals from all three layers. Firmographic fit might contribute forty percent of the score — does this company match your ICP's size, industry, technology, and growth characteristics? The person layer contributes another thirty percent — have you identified the right decision-maker, do you have verified contact information, and do they match the profile that historically converts? Timing signals contribute the remaining thirty percent — are there active indicators that this company is ready to buy right now?

The weights aren't arbitrary. They come from analyzing your historical deal data. Which signals were present when your best customers bought? Which signals were present when deals stalled or lost? Every company's scoring model is different because every company's winning patterns are different. The ICP Blueprint you built in Chapter 3 — specifically the customer analysis and deal pattern work — gives you the inputs to calibrate your model.

The output is a ranked priority list that refreshes weekly. Company A might score one forty-five this week because of a fresh funding round plus active hiring plus a new CRO. Company B might score ninety-two — a solid fit with no urgent timing signals. But next week, Company B suddenly posts seven sales roles and their CRO leaves. Their score jumps to one thirty-eight. Without the weekly refresh, you'd have missed that window entirely.

Integrating intelligence with your CRM. Intelligence data that lives in a separate platform is intelligence data that nobody uses. The critical step is mapping enrichment fields into your CRM so that lead scoring, routing, and personalization happen automatically.

This means creating custom fields in your CRM for every intelligence dimension that matters to your scoring model — ICP fit score, technology stack, recent signals, hiring status, funding status, enrichment date, and data source. It means configuring automated workflows that route high-scoring prospects to the right reps, trigger personalized sequences based on specific signal combinations, and flag accounts that cross scoring thresholds. And it means establishing a data refresh cadence — weekly for your active pipeline, monthly for your broader database — so that your intelligence doesn't go stale.

The integration architecture matters because it determines whether your sales team actually uses the intelligence or ignores it. If enrichment data lives in a separate tab in Clay or ZoomInfo that reps have to manually check, adoption will be twenty percent at best. If enrichment data populates the fields your reps already look at in their CRM — the lead record, the account page, the activity feed — adoption approaches one hundred percent because the intelligence is embedded in their existing workflow.

Reading the Dashboard: How Leaders Make Decisions from Intelligence Data

Data without interpretation is noise. Once your intelligence system is running, the challenge shifts from "we don't have enough data" to "we have too much data and don't know what matters." Executive dashboards solve this by organizing intelligence into three levels of decision-making.

At the strategic level, you're asking big questions. What's our total addressable market within our ICP? What percentage have we penetrated? Where are the whitespace opportunities — segments or geographies where we have strong fit but low penetration? Is our ICP definition still accurate, or are our best-performing deals coming from a segment we didn't originally target? These are quarterly conversations that shape resource allocation and go-to-market strategy.

At the operational level, you're asking execution questions. Which audience segments are converting best? Where are deals stalling in the pipeline? What's our cost per meeting by intelligence source — are the prospects from intent data converting at a higher rate than the ones from firmographic targeting alone? Is our scoring model accurate, or are low-scored prospects outperforming high-scored ones? These are weekly conversations that drive tactical adjustments.

At the tactical level, you're asking daily questions. Which accounts showed new buying signals this week? Who on the Top 25 list has responded? Which reps are leveraging enrichment data effectively and which ones are ignoring it? What messages are resonating with which segments? These are the conversations that happen in Monday standups and Friday reviews.

The discipline of reviewing intelligence data at all three levels is what separates companies that get compound returns from their intelligence investment from companies that buy expensive platforms and use ten percent of the capability.

Competitive Intelligence: The Layer Most Companies Ignore

There's a dimension of audience intelligence that doesn't fit neatly into the three-layer framework but is too valuable to skip: competitive intelligence. Understanding what your competitors are doing — their positioning, their pricing, their wins and losses — directly impacts how you approach every prospect in your pipeline.

AI-powered competitive analysis monitors competitor mentions across social media, news, analyst reports, and customer reviews in real time. It identifies patterns in your win/loss data that reveal exactly where you beat competitors and where they beat you. It tracks competitor funding, acquisitions, and product launches that might shift the competitive landscape. And it feeds directly into your sales enablement — battle cards that update automatically, objection-handling frameworks based on what prospects actually say in competitive evaluations, and differentiation messaging tuned to the specific competitors each prospect is considering.

The companies that use competitive intelligence well don't just respond to competitors — they anticipate them. If AI monitoring detects that a competitor just raised a large round and is hiring aggressively in your core market, you adjust your positioning and outreach cadence before their new campaigns hit. If win/loss analysis reveals that you consistently lose deals where a specific competitor is involved and the CFO is the primary decision-maker, you develop targeted messaging for exactly that scenario.

Competitive intelligence is the difference between playing defense — reacting when a prospect mentions a competitor in a call — and playing offense — knowing which competitors each prospect is evaluating before the first conversation and having your differentiation ready.

Building Your Intelligence System: The Practical Path

The goal of this chapter isn't to make you an expert in every intelligence platform. It's to help you understand what's possible, make the right platform decisions, and build a system that your team will actually use.

The practical path follows three phases. In the first two weeks, you're building the data foundation — selecting platforms, integrating them with your CRM, configuring enrichment workflows, and establishing data quality standards. The decisions you make here about which platforms to use, how they connect, and what data quality thresholds you'll enforce set the ceiling for everything that follows.

In weeks three and four, you're building the intelligence layer itself — creating your scoring model based on historical deal data, segmenting your audience into intelligence-based tiers, establishing routing rules that get the right prospects to the right reps, and developing personalization rules that translate intelligence data into specific messaging strategies for each segment.

In weeks five through eight, you're activating and optimizing. Deploy intelligence-driven campaigns, run A/B tests comparing intelligence-targeted outreach against your previous generic approach, monitor performance by segment, collect feedback from your sales team about what's working and what isn't, and refine your scoring models based on actual conversion data. By the end of this phase, you'll have hard numbers showing the impact of intelligence on response rates, meeting conversion, pipeline velocity, and cost per acquisition.

The companies that do this well see dramatic improvements. Response rates climb from the four to six percent range into the mid-teens. Meeting conversion rates double or triple. Sales cycles shorten because you're reaching the right people at the right time with the right message. And customer acquisition costs drop because you stop spending money on prospects who were never going to buy.

Chapter 4 Executive Takeaways

For the CEO: Audience intelligence transforms AI sales from spray-and-pray to precision targeting. The three-layer model — company, person, and timing — is what makes the difference between generating activity and generating revenue. The investment in intelligence platforms pays for itself through improved conversion rates and reduced customer acquisition costs. This is where your ICP Blueprint from Chapter 3 becomes operational.

For the CRO: Intelligence-driven targeting enables sales teams to focus on high-probability prospects with personalized messaging. The scoring model and Top 25 methodology give your team a systematic framework for prioritization instead of gut-feel prospecting. When your reps have enriched data on every prospect — company context, technology stack, recent signals, and AI-generated personalization hooks — their conversations start at a fundamentally different level.

For the CMO: Audience intelligence bridges the gap between marketing and sales by providing shared definitions of target prospects and engagement strategies. Intent signals tell you which accounts are actively researching your category, allowing marketing to coordinate air cover with sales outreach. This alignment is where the "marketing generates leads that sales doesn't want" problem finally gets solved.

For the VP of Sales: Intelligence platforms provide sales reps with the context and insights needed for consultative selling. Reps armed with firmographic, technographic, and intent data consistently outperform those using generic approaches — not because the reps are better, but because they're spending their time on the right prospects with the right information at the right moment.

Preparing for Blueprint Creation

With your ICP architecture defined and audience intelligence system operational, you're ready for the final component of the Architecture phase: creating your boardroom-grade ICP Blueprint.

In Chapter 5, we'll transform your persona research and intelligence systems into a comprehensive document that guides all AI tool selection, integration, and optimization decisions.

Your architecture is nearly complete. The blueprint makes it actionable.

References:

Bombora. (2024). *The state of intent data: 2024 benchmark report*. Bombora Company Surge Research.

6sense. (2024). *Revenue AI benchmark report: B2B buyer behavior and predictive intelligence trends*. 6sense Revenue AI Research.

ZoomInfo. (2024). *The state of pipeline generation: Data-driven prospecting benchmark study*. ZoomInfo Research Division.

CHAPTER 5

Building Your Blueprint

I've read hundreds of sales strategy documents. Pitch decks, go-to-market plans, persona playbooks, "strategic frameworks" that cost six figures from brand-name consulting firms. Most of them are the same forty pages of well-formatted platitudes: "Target mid-market companies." "Focus on value-based selling." "Build deeper relationships." "Leverage our competitive advantages."

They look great in a board presentation. They're completely useless on a Monday morning.

Ask the sales rep sitting in front of a blank email at 8 AM what to do with that document and watch what happens. They can't tell you which ten companies to call this week. They can't explain why those companies and not the other two hundred on the list. They can't articulate what pain point to lead with for a CFO versus a VP of Operations. They can't describe the trigger event that signals this prospect is ready to buy right now instead of six months from now.

So they do what every rep does when the strategy doc doesn't help - they wing it. They fall back on gut instinct, personal relationships, and whatever worked last quarter. The $150,000 go-to-market deck sits in a shared drive that nobody opens after the first week.

Now imagine a different document. One page. It tells your sales team exactly who to target - not an industry and a company size, but a specific profile with qualification criteria they can apply in sixty seconds. It tells your marketing team exactly what to say - messaging mapped to each persona's role, pain points, and decision-making triggers. It tells your AI systems exactly how to score, prioritize, and sequence - because the parameters are specific enough to configure, not vague enough to interpret.

And it tells your board exactly why this will work - because the ICP is tied to historical win data, market sizing, and revenue projections that connect targeting decisions to financial outcomes.

That document is the ICP Blueprint. It's the deliverable from everything you built in Chapters 3 and 4 - the alignment workshop, the persona discovery, the three intelligence layers - distilled into a single strategic artifact that drives every decision from this point forward. Not a persona deck. Not a slide. A blueprint, in the architectural sense - specific enough to build from, clear enough for anyone on the team to read, and precise enough that your AI systems can execute against it without guessing.

The difference between companies that "use AI tools" and companies that have AI Sales Engines comes down to whether this document exists. If it does, everything downstream - tool selection, data enrichment, campaign design, optimization - has a foundation. If it doesn't, you're just buying power tools without a plan again.

Producing a Boardroom-Grade ICP Blueprint Document

A boardroom-grade ICP Blueprint meets four critical requirements that distinguish it from typical marketing personas or sales playbooks. First, it demonstrates strategic clarity with a clear connection between ICP and overall business strategy. Second, it provides operational precision that's specific enough to guide AI tool configuration and sales execution. Third, it maintains executive accessibility - understandable and actionable for C-level decision makers who may not be involved in day-to-day sales activities. Fourth, it ensures implementation readiness by containing everything needed to begin AI system assembly without additional research or analysis.

The blueprint structure follows a systematic format that's been refined through dozens of implementations. The **Executive Summary** spans two pages and covers strategic rationale for ICP selection, expected business impact and ROI projections, implementation timeline and resource requirements, and success metrics and measurement methodology.

Section 1: Market Analysis and Opportunity Assessment requires four to six pages covering Total Addressable Market (TAM) analysis by ICP segment, competitive landscape and positioning analysis, market growth projections and trend analysis, and strategic fit with company capabilities and objectives.

Section 2: Ideal Customer Profile Definition spans eight to ten pages detailing primary ICP characteristics and qualification criteria; secondary ICP segments and prioritization framework; firmographic, technographic, and behavioral specifications; and geographic and market segment parameters.

Section 3: Buyer Persona Profiles covers six to eight pages with decision maker profiles including roles, responsibilities, and motivations; influencer mapping and stakeholder analysis; buying process documentation and timeline analysis; and communication preferences and engagement strategies.

Section 4: Value Proposition and Messaging Framework requires four to six pages covering primary value propositions for each stakeholder; competitive differentiation and positioning; messaging hierarchy and supporting evidence; and objection handling and competitive responses.

Section 5: Go-to-Market Strategy spans six to eight pages detailing channel strategy and resource allocation, campaign strategies for each ICP segment, sales process alignment and methodology, and a marketing and sales coordination framework.

Section 6: Implementation Plan and Success Metrics

covers four to six pages including AI tool selection criteria based on ICP requirements, integration architecture and data requirements, success metrics and measurement methodology, and implementation timeline and milestone tracking.

The document quality standards ensure professional presentation with executive-level design and formatting; data visualization for key insights and trends; professional charts, graphs, and infographics; and comprehensive appendices with supporting data. Content accuracy standards require all data sources cited and verified, statistical analysis reviewed by qualified analysts, customer insights validated through interviews, and market research verified through multiple sources.

The stakeholder review process includes individual stakeholder review and feedback incorporation, group review session with all key participants, final approval from the CEO and executive team, and version control and update process establishment.

How Precision ICP Strategy Transforms Results

The pattern is consistent across every industry and business model I've worked in: companies that move from generic targeting to precision ICP Blueprints see dramatic improvements across every sales metric.

The transformation follows a predictable arc. Before the Blueprint process, companies target broad categories - "mid-market companies that need better project management" or "enterprise organizations interested in cybersecurity." Their messaging is generic, their response rates are low, and their sales cycles are long because they're having the wrong conversations with the wrong people about the wrong problems.

After the Blueprint process, they target precise segments with specific firmographic, technographic, and behavioral characteristics. Their messaging resonates because it speaks to specific challenges that their prospect actually faces. Response rates jump dramatically - typically tripling or more - because prospects feel understood rather than targeted. Sales cycles compress because conversations start at a higher level of relevance, and deal values increase because the value proposition maps directly to the prospect's situation.

The three most common Blueprint transformation patterns illustrate this:

Vertical Precision. Companies that move from industry-agnostic targeting to vertical-specific ICPs consistently see the largest improvements. When a SaaS company stops targeting "all mid-market companies" and starts targeting "architecture and engineering firms with distributed teams, regulatory compliance requirements, and legacy project management tools," every metric transforms. The messaging can reference specific pain points, the value proposition can emphasize industry-specific ROI, and the enrichment data can identify precise buying triggers like new project wins, regulatory changes, and team expansion.

Account-Based Alignment. Enterprise companies that move from broad Fortune 1000 targeting to buying-committee-aligned ICP strategies unlock deals that were previously unwinnable. The Blueprint process maps every stakeholder in the buying committee - executive sponsor, technical evaluator, financial approver, end-user champion, compliance validator - and develops role-specific value propositions for each. This multi-threaded approach transforms win rates and dramatically reduces customer acquisition costs because you're engaging the entire decision-making unit rather than trying to push a deal through a single contact.

Founder Independence. Growth-stage companies that move from founder-dependent relationship selling to systematic ICP-driven processes unlock scalability. The Blueprint captures the founder's intuitive understanding of ideal customers in documented, replicable criteria that any rep can use. The result is consistent pipeline generation that doesn't depend on any individual's relationships, with deals progressing systematically rather than relying on personal networks.

In every case, the improvement comes from the same source: replacing broad assumptions with precise, data-validated targeting criteria that the entire sales organization can execute consistently.

The Ultimate Blueprint Checklist

Creating a boardroom-grade ICP Blueprint requires systematic attention to detail and comprehensive coverage of all strategic and operational elements. This checklist ensures your blueprint meets professional standards and implementation requirements.

Strategic Foundation Checklist

Market analysis components must include Total Addressable Market (TAM) calculation with methodology, Serviceable Addressable Market (SAM) analysis, competitive landscape mapping with positioning analysis, market growth trends and future projections, and regulatory and industry trend impact analysis.

Business case development requires revenue projections by ICP segment, customer acquisition cost analysis and targets, customer lifetime value calculations, market share objectives and growth trajectory, and resource allocation and investment requirements.

ICP Definition Checklist

Primary ICP characteristics must specify firmographic specifications (size, industry, geography), technographic requirements (current stack, integration needs), behavioral indicators (growth stage, buying patterns), financial parameters (budget, procurement process), and organizational structure (decision-making hierarchy).

Secondary ICP segments require alternative ICP definitions with prioritization criteria, market size and opportunity assessment for each segment, resource allocation strategy across segments, and success metrics and measurement approach.

Buyer Persona Checklist

Decision maker profiles must include job titles, levels, and reporting relationships; key responsibilities and performance metrics; pain points and business challenges; success criteria and personal motivations; and communication preferences and timing.

Buying process documentation requires decision journey stages and timeframes, stakeholder involvement and influence patterns, evaluation criteria and selection process, budget approval and procurement requirements, and implementation planning and timeline expectations.

Messaging Framework Checklist

Value proposition development must include primary value proposition with supporting evidence, stakeholder-specific messaging variations, competitive differentiation and positioning, ROI calculations and business case support, and risk mitigation and implementation advantages.

Content strategy alignment requires content requirements for each buying stage, channel preferences and engagement strategies, personalization parameters for AI implementation, success stories and proof points, and objection handling and competitive responses.

Implementation Readiness Checklist

AI tool selection criteria must specify functional requirements based on ICP characteristics, integration requirements for existing systems, data quality and enrichment needs, personalization and automation capabilities, and scalability and performance requirements.

Success measurement framework requires Key Performance Indicators (KPIs) definition, measurement methodology and reporting schedule, attribution model for revenue tracking, data quality standards and validation process, and optimization and improvement process.

Blueprint Validation and Stakeholder Alignment

The validation process ensures your blueprint accurately reflects market reality and organizational capabilities, while maintaining stakeholder commitment to implementation.

Internal Validation Process

The executive review session requires two hours to present the complete blueprint to C-level stakeholders, address questions and concerns about strategy and implementation, secure formal approval and resource commitment, and establish success metrics and accountability framework.

Departmental alignment reviews ensure cross-functional support. Sales team validation involves reviewing ICP characteristics for practical targeting feasibility, affirming messaging frameworks for conversation effectiveness, assessing persona accuracy based on field experience, and confirming success metrics and measurement approach.

Marketing team alignment includes aligning content strategy with ICP and persona requirements, coordinating channel strategy and campaign development, integrating lead scoring and qualification criteria, and establishing attribution and performance tracking.

Customer success integration involves validating ICP characteristics against successful customer patterns, aligning expansion and retention strategies with ICP framework, integrating customer feedback into persona refinement, and coordinating reference and case study development.

External Validation Methods

Customer advisory board review involves presenting ICP and persona insights to key customers, validating buying process documentation and timeline accuracy, collecting feedback on value proposition relevance and messaging, and refining strategies based on customer input.

Industry expert consultation includes reviewing blueprint with industry analysts and consultants, validating market opportunity and competitive positioning, incorporating industry trend analysis and future projections, and benchmarking against industry best practices.

Pilot campaign testing involves launching limited-scale campaigns targeting defined ICPs, measuring response rates and engagement metrics, collecting qualitative feedback from prospects and sales team, and refining the approach based on real-world performance data.

Blueprint Evolution and Continuous Improvement

Your ICP Blueprint is a living document that must evolve with market changes, competitive dynamics, and business growth. The evolution process ensures your blueprint remains accurate and effective over time.

Quarterly Review Process

Performance analysis examines achievement against success metrics, pipeline quality and conversion rates, message resonance and engagement effectiveness, and competitive positioning and market changes.

Strategy refinement involves updating ICP characteristics based on performance data, refining messaging based on customer feedback and results, adjusting targeting parameters for improved accuracy, and optimizing resource allocation across ICP segments.

Market intelligence integration includes incorporating new market research and competitive intelligence, updating opportunity assessments and growth projections, refining positioning based on industry trends, and adapting strategy for regulatory or technology changes.

Annual Blueprint Refresh

Comprehensive market analysis requires complete TAM/SAM recalculation, competitive landscape reassessment, customer base analysis and segmentation review, and technology trend impact analysis.

Strategic realignment involves business strategy alignment verification, resource allocation and investment planning, success criteria and measurement framework updates, and long-term growth strategy integration.

Chapter 5 Executive Takeaways

For the CEO: The ICP Blueprint is your strategic roadmap for AI sales transformation. It connects your AI investment to specific business outcomes and provides the foundation for scalable revenue growth. Without this blueprint, AI implementations become expensive experiments rather than strategic investments.

For the CRO: A comprehensive blueprint eliminates sales team confusion and provides clear targeting and messaging guidance. This clarity directly improves conversion rates and sales efficiency while enabling systematic scaling and performance management.

For the CMO: The blueprint ensures sales and marketing alignment around shared definitions of ideal customers and messaging strategies. This alignment is essential for integrated AI sales and marketing systems that generate consistent results.

For the Board: The ICP Blueprint demonstrates strategic thinking and systematic approach to market opportunity. It provides confidence that AI investments will generate measurable business results and supports valuation enhancement through demonstrated systematic revenue generation.

Transitioning from Architecture to Assembly

With your ICP Blueprint complete, validated, and approved, you've finished the Architecture phase of the AI Sales Engine methodology. You now have complete stakeholder alignment on target customer definition, comprehensive understanding of buyer personas and decision processes, clear messaging frameworks for each stakeholder and buying stage, validated market opportunity and competitive positioning, and success metrics and measurement methodology.

You're ready to move to Phase 2: Assembly.

In the Assembly phase, you'll transform your strategic architecture into operational capability through AI tool selection, system integration, and process development. Your blueprint becomes your guide for every technology decision and integration choice. The foundation is set. Now we build the engine that will generate predictable revenue growth and sustainable competitive advantage.

PHASE 2: ASSEMBLE

CHAPTER 6

Selecting the Right AI Stack

Companies evaluating AI sales tools end up in one of two places. Neither is good.

The first is paralysis. You start researching platforms, and within a week you've got a spreadsheet with thirty vendors, each claiming to be the only tool you'll ever need. Every demo looks impressive. Every case study shows triple-digit improvements. Every salesperson tells you that their competitor's product is overpriced garbage. You schedule more demos. You read more reviews. You build a comparison matrix that gets more confusing every time you add a column. Three months later, you've evaluated everything and bought nothing. Your team is still doing things manually while you're stuck in an endless loop of "let's see one more demo."

The second is chaos. You buy everything. The CEO sees a prospecting tool at a conference - buy it. The CRO's friend swears by a conversation intelligence platform - buy it. Marketing finds an email automation tool that integrates with HubSpot - buy it. Someone on the team signs up for three different AI writing assistants on free trials and now all of them are auto-renewing at $500 a month. Twelve months later, you're spending $40,000 a month on platforms that don't talk to each other, your sales team is toggling between eight tabs all day, and nobody can explain which tool is actually responsible for the pipeline you are generating.

This chapter is about a third option: selecting the right AI stack based on what your business actually needs, what integrates with what you already have, and what serves the system you're building - not what looked good in a demo.

The blueprint you built in Chapter 5 makes this possible. Without it, tool selection is a guessing game - which is how you end up with paralysis or chaos. With it, you have specific criteria for every decision. Does this tool serve our ICP? Does it integrate with our CRM? Does it solve a problem we've actually identified, or is it a solution looking for a problem? Does it fit the Architect-Assemble-Activate sequence, or does it skip steps?

Before you buy a single new tool, we're going to start somewhere most companies skip entirely - a clear-eyed assessment of what you already have.

Start With What You Already Have: The Current State Assessment

Before you buy a single new AI tool, you need to understand your existing technology foundation. I've seen companies spend fifty thousand dollars on AI platforms only to discover they don't integrate with their current CRM. Or that their data quality is so bad the AI can't do anything useful with it. Or that their sales process is so broken that automating it just scales the brokenness.

The quality of your existing infrastructure determines the success or failure of AI implementation more than the sophistication of the AI tools themselves.

Here's how to assess what you're starting with.

First, look at your CRM platform. What are you using - Salesforce, HubSpot, Pipedrive, something else? What tier or edition - Professional, Enterprise, something in between? How long have you been using it? What's your data quality like - can you trust the information in there or is it a mess?

The CRM is your system of record. Everything else will connect to it. If your CRM is a dumpster fire, no amount of AI will fix that. You need to clean it up first.

Second, evaluate your current sales process. Do you have documented processes or is everyone doing their own thing? Are your processes consistent across the team or does every rep have their own approach? What's actually working versus what's theoretical? Where are the obvious bottlenecks that slow everything down?

AI amplifies your process. If your process is broken, AI makes it broken faster.

Third, assess your technical infrastructure. What's your data architecture like - where does data live and how does it flow? Do you have API access to your current tools? What integration tools are you using - Zapier, Make, native integrations? Do you have someone technical on staff or are you entirely non-technical?

You don't need to be a tech company to implement AI, but you do need to understand your technical constraints.

Fourth, look at your marketing technology stack. What marketing automation are you using if any? How do marketing and sales currently share data? What analytics and reporting tools do you have? Are marketing and sales aligned or operating in silos?

AI sales tools work best when they're connected to your marketing systems. If there's a wall between marketing and sales, AI can't fix that fundamental misalignment.

Fifth, examine your analytics and reporting. Can you currently track the sales funnel accurately? Do you know your conversion rates at each stage? Can you attribute revenue to specific activities? What's your forecast accuracy like?

AI gives you better data and insights, but only if you have the foundation to capture and measure them.

Finally, evaluate your integration readiness. Do your current tools talk to each other or are they islands? Can you move data between systems easily or is everything manual? Do you have documented data flows or is it all tribal knowledge? Are there obvious integration gaps that will cause problems?

Integration is where most AI implementations fail. Not because the AI doesn't work, but because it can't talk to the rest of your stack.

I had a client - forty-three million in revenue, twenty-person sales team - who wanted to implement conversation intelligence AI. They were using Salesforce as their CRM. Seemed straightforward. Then we discovered they were on Salesforce Professional edition, which didn't have the API access needed for real-time call recording integration. Their options were upgrade to Enterprise at an extra eighteen thousand per year or manually upload call recordings, which defeated the whole purpose of automation. They hadn't budgeted for the upgrade. The entire project stalled for three months while they figured out funding.

That's a current state assessment problem, not a tool selection problem. We could have avoided it entirely by checking the technical requirements first.

Here's the hard truth: if your assessment reveals major gaps - terrible data quality, no documented processes, tools that don't integrate, misalignment between teams - you need to fix those before you buy AI tools. I know that's not what you want to hear. You want the AI magic now. But building AI on a broken foundation is like building a mansion on quicksand.

Fix the foundation first. Then add the AI.

The Five Essential Categories: What Your AI Sales Engine Actually Needs

Based on working with over two hundred companies implementing AI sales systems, I can tell you that successful AI Sales Engines require five core categories of AI functionality.

The key is understanding that you don't need every possible AI capability that exists. You need the right capabilities that serve your specific ICP Blueprint and sales process.

Let me walk you through each category, the tools that matter, and how to think about them.

Category 1: Intelligent Prospecting and Lead Generation

This is the foundation. You need tools that help you find the right people, enrich their data, and enable personalized outreach at scale.

Clay.com has become our go-to recommendation for most companies in the five to one hundred million dollar range, and I'm going to spend more time on this one because it's different from traditional data platforms.

Clay costs three forty-nine to eight hundred dollars per month per seat, plus usage-based enrichment credits. Typical total cost runs five hundred to two thousand per month for a mid-market company. The enrichment credits themselves range from point-zero-zero-one to fifty cents per enrichment depending on which data source you're using.

What makes Clay different is the waterfall enrichment model. Traditional platforms like ZoomInfo give you their data or nothing. Clay connects to over fifty data providers and lets you build workflows that try multiple sources in sequence. Try the cheap source first, if it fails try the mid-price source, if that fails try the expensive source. You only pay for successful enrichments. This gives you eighty-five to ninety-five percent success rates instead of forty to sixty percent, and it costs a fraction of what you'd pay for a ZoomInfo enterprise contract.

Clay also has native GPT integration, which means you can use AI to analyze enrichment data and generate personalized messaging during the enrichment process itself. Your reps don't just get data - they get ready-to-use personalization hooks.

Clay is best for companies that need enrichment flexibility without enterprise contracts, want to test targeting strategies before committing big budgets, need custom workflows built to their specific ICP, or want AI personalization built into the enrichment process.

The learning curve is moderate. If you're comfortable with spreadsheets and logical thinking, you can learn Clay. It's not coding, but it's not brain-dead simple either. Implementation takes two to four weeks typically.

When NOT to choose Clay: If you need fully managed service with guaranteed SLAs and someone to call when things break, go with ZoomInfo. If you want completely turnkey simplicity with zero workflow building, use Apollo. If you're enterprise scale over one hundred million and have dedicated ops teams, ZoomInfo might be worth the premium.

ZoomInfo SalesOS is still the gold standard for enterprise companies, despite the cost. Pricing runs fourteen thousand nine hundred ninety-five to twenty-four thousand nine hundred ninety-five dollars annually for mid-market teams, but enterprise contracts can hit fifty to one hundred thousand.

What you get is the most comprehensive prospect database with AI-enhanced searching, intent data integration showing who's actively researching your category, buying signal identification that flags companies entering buying mode, automated outreach sequences with some personalization, and deep CRM integration that works natively with Salesforce and HubSpot.

ZoomInfo is best for companies over one hundred million in revenue with fifty-plus person sales teams who need managed service, guaranteed ninety-five percent plus data accuracy, dedicated support and training, and compliance infrastructure for regulated industries.

The trade-off is cost and flexibility. You're paying for enterprise-grade accuracy and support, but you're locked into their data model and annual contracts.

Apollo.io sits in the middle as the budget-friendly all-in-one option. Pricing is forty-eight hundred to ninety-six hundred annually for similar functionality to ZoomInfo but at a fifth of the cost.

Apollo provides AI-powered prospecting with technographic data, email finder and verification capabilities, sequence automation with A/B testing built in, and analytics and performance optimization all in one platform.

Apollo is best for teams of two to ten people who want simplicity over optimization, need to get started quickly without a lot of setup, have limited budget at five to fifteen thousand total per year, and can accept seventy to eighty percent enrichment rates instead of ninety-plus.

The downside is less flexibility than Clay and less accuracy than ZoomInfo. But if you need to prove the model works before investing heavily, Apollo is a solid starting point.

Outreach.io is worth mentioning in this category even though it's really a sequence platform. Pricing is seventy-two hundred to sixteen thousand eight hundred dollars annually per user.

Outreach provides AI-powered sequence optimization and personalization, conversation intelligence and coaching insights, advanced analytics and performance tracking, and enterprise-grade security and compliance.

Outreach makes sense for larger teams that already have their data sourcing figured out and need sophisticated sequence automation. If you're using Clay or ZoomInfo for data and just need the outreach layer, Outreach is powerful.

When you're evaluating tools in this category, your selection criteria should focus on these factors. Database quality and coverage - do they have the contacts you need? Integration capabilities - will this actually talk to your CRM? AI personalization depth - is it truly intelligent or just mail merge? Compliance and data privacy - will this pass your legal team's review? Total cost of ownership including setup time, training needs, and ongoing costs beyond the subscription.

Let me give you a real decision framework example. Say you're a thirty million dollar B2B SaaS company with eight sales reps targeting mid-market accounts. You have HubSpot as your CRM. Your budget for prospecting and enrichment is fifteen thousand per year.

ZoomInfo is overkill and over-budget. Their minimum would eat your entire budget and you don't need enterprise SLAs. Apollo would work and stay in budget, but you'd get seventy percent enrichment rates and basic personalization. Clay plus a lighter Apollo subscription would give you ninety percent enrichment through waterfalls, AI-generated personalization hooks, and stay under budget at twelve to fourteen thousand per year total.

For that scenario, Clay plus Apollo is the right answer. But if your data quality requirements were over ninety-five percent because you're in healthcare or finance, ZoomInfo becomes worth the premium despite the cost.

There's no universal "best" tool. There's only the best tool for your specific situation.

Category 2: Conversational AI and Communication Enhancement

This category is about understanding what's happening in your sales conversations and using AI to make those conversations better.

Gong.io is the category leader here. Pricing runs twelve thousand to nineteen thousand two hundred dollars annually per user, which sounds insane until you see what it does.

Gong records all your sales calls and uses AI to analyze them. Not just transcription - actual analysis. It identifies which parts of your pitch are working and which are falling flat. It spots when deals are at risk based on conversation patterns. It gives you pipeline forecasting based on deal health. It provides coaching recommendations showing exactly what each rep needs to improve. And it integrates with major CRM and communication platforms so everything flows automatically.

Gong is expensive because it's not just a tool - it's a coaching system, a forecasting engine, and a conversation intelligence platform all in one. For companies where deal sizes justify the investment, it's transformative.

Best for: teams where average deal size is fifty thousand plus, sales managers who need to coach at scale, companies with complex sales processes that need visibility, and organizations serious about sales performance improvement.

Chorus.ai used to be Gong's main competitor but got acquired by ZoomInfo. If you're already a ZoomInfo customer, Chorus pricing is included in your ZoomInfo contract, which makes it essentially free if you're already paying for ZoomInfo.

Chorus provides call recording and conversation analysis, competitive intelligence pulled from calls, deal acceleration recommendations, and team performance analytics.

Honestly, if you have ZoomInfo, use Chorus. If you don't have ZoomInfo, Gong is probably better because it's more mature and has better AI analysis.

Otter.ai for Business is the budget option at twelve hundred to twenty-four hundred dollars annually per user. It's not as sophisticated as Gong, but it covers the basics.

You get meeting transcription that's quite accurate, AI-generated summaries of calls, action item extraction so nothing falls through cracks, and CRM integration to log everything automatically.

Otter makes sense for smaller teams that need transcription and basic call logging but don't need sophisticated coaching analytics. If your average deal size is under twenty thousand dollars, Gong's cost probably doesn't make sense. Start with Otter.

The decision in this category usually comes down to deal economics. Take your average deal size, multiply by your close rate, divide by your average number of touches to close. That gives you revenue per sales activity. If Gong's cost is less than ten percent of that number, the ROI is probably there. If not, start with something lighter.

Category 3: Predictive Analytics and Pipeline Intelligence

This is where AI helps you see around corners - predicting which deals will close, which leads are worth pursuing, and where to focus your time.

Salesforce Einstein Analytics is the obvious choice if you're already on Salesforce. Pricing is twenty-four hundred to thirty-six hundred dollars annually per user, which gets added on top of your Salesforce subscription.

You get native CRM integration that just works because it's built by Salesforce, predictive lead scoring based on your historical data, AI-generated insights that surface automatically, and the ability to build custom AI models for your specific needs.

The huge advantage is it lives inside Salesforce. Your reps don't have to switch tools. The data is already there. The predictions appear right in the opportunity records.

The disadvantage is you're locked into the Salesforce ecosystem. If you're on HubSpot or something else, this doesn't help you.

6sense Revenue AI is the enterprise account-based platform. Pricing runs eighteen thousand to thirty-six thousand dollars annually, which is expensive but powerful.

Six sense does account-based predictive analytics showing which accounts are in-market, intent monitoring that tracks which accounts are researching your category, pipeline acceleration recommendations, and market intelligence about competitive activity.

Six sense makes sense when you're doing true account-based selling with deal sizes over one hundred thousand dollars, you have long sales cycles where timing intelligence matters, you need to coordinate across marketing and sales teams, and you can afford the enterprise price tag.

People.ai focuses on activity capture and revenue intelligence. Pricing is ninety-six hundred to fourteen thousand four hundred dollars annually per user.

People automatically captures all sales activity from email, calendar, and calls so reps don't have to log everything manually. It provides revenue intelligence and forecasting based on actual activity patterns. It shows performance optimization opportunities by comparing top performers to others. And it gives you visibility into who's doing what across the entire team.

People is best for sales leaders who want to understand activity patterns, companies where CRM data entry is a constant battle, organizations that need accurate forecasting, and teams where you want to identify what top performers do differently.

When evaluating this category, the key question is what problem you're actually trying to solve. If you need better lead scoring, Einstein or a CRM-native solution might be enough. If you need account-based intelligence, six sense is probably worth it. If you need activity visibility and coaching data, People makes sense.

Don't buy predictive analytics because it sounds cool. Buy it because you have a specific forecasting or prioritization problem it will solve.

Category 4: Content Intelligence and Personalization

This category is about making sure your sales content is actually effective and gets used appropriately.

Seismic is the enterprise standard. Pricing runs fourteen thousand four hundred to twenty-one thousand six hundred dollars annually per user.

Seismic provides content management and optimization so reps can actually find the right content, sales enablement and training so they know how to use it, analytics showing what content drives results, and CRM integration so content usage is tracked.

Seismic makes sense for enterprise teams with large content libraries, companies where content quality and compliance matter, organizations that need detailed content analytics, and teams that struggle with content discoverability.

Showpad is similar to Seismic with slightly different strengths. Pricing is twelve thousand to eighteen thousand dollars annually per user.

You get content intelligence and analytics, AI-powered content recommendations showing what works best in different situations, training and coaching built around content, and sales engagement tracking.

Showpad tends to be a bit more user-friendly than Seismic. If your team is less technical, Showpad might be easier to adopt.

Highspot rounds out the category at thirteen thousand two hundred to nineteen thousand two hundred dollars annually per user.

Highspot focuses on content management, sales playbook integration so content aligns with process, analytics and AI-powered insights, and buyer engagement tracking.

Honest talk: these three tools - Seismic, Showpad, Highspot - are all good. They do roughly the same things with slightly different interfaces and feature prioritization. Your decision here is more about which UI your team prefers and which integrates best with your specific CRM and sales tools.

Most companies under fifty million in revenue don't need dedicated content intelligence platforms. If you're smaller, your CRM's content management plus a shared drive is probably enough. Save the fifteen thousand per year until content management becomes a real bottleneck.

Category 5: Marketing and Sales Alignment

The last category is about connecting marketing and sales so leads flow smoothly and both teams work from the same data.

HubSpot Sales Hub Enterprise is my default recommendation for most companies between five and fifty million in revenue. Pricing is around fourteen thousand four hundred dollars annually for the platform, which covers multiple users depending on your specific setup.

You get integrated CRM, marketing automation, and sales tools all in one ecosystem. AI-powered lead scoring and routing happen automatically. Workflow automation connects marketing activities to sales follow-up. And analytics span both marketing and sales so you can see the full funnel.

The massive advantage of HubSpot is simplicity. It's one platform. Marketing and sales live in the same system. Data doesn't need to sync between tools. AI features are baked in. And the learning curve is manageable even for non-technical teams.

Best for: companies that don't currently have marketing automation, teams that want simplicity over best-of-breed complexity, organizations where marketing and sales alignment is a problem, and companies with budgets under fifty thousand per year for the whole stack.

Marketo Engage is the enterprise alternative for larger companies or those with complex needs. Pricing runs eighteen thousand to thirty-six thousand dollars annually.

Marketo provides sophisticated marketing automation, advanced lead nurturing and scoring, account-based marketing capabilities, and multi-touch attribution showing what actually drives revenue.

Marketo makes sense when you're over fifty million in revenue with dedicated marketing ops resources, you have complex lead scoring and routing needs, you're doing serious account-based marketing, and you need detailed attribution modeling.

The downside is complexity. Marketo is powerful but hard to use. You really need a dedicated Marketo admin. For smaller companies, it's overkill.

Pardot from Salesforce rounds out the category at fifteen thousand to twenty-four thousand dollars annually.

Pardot gives you native Salesforce integration that just works, B2B marketing automation built for sales teams, lead scoring and grading, and ROI reporting that ties back to opportunities.

Pardot makes sense if you're already deep into Salesforce and want the marketing automation layer to match. The integration is tighter than connecting Marketo to Salesforce, and it's built specifically for B2B sales motions.

When you're evaluating this category, the big question is whether you want integrated simplicity or best-of-breed complexity. HubSpot gives you good-enough functionality across marketing and sales in one simple platform. Marketo or Pardot plus Salesforce gives you more power but more complexity.

For most companies reading this book, HubSpot is the right answer until you hit fifty million in revenue. Then you can reassess whether you need to graduate to Salesforce plus Marketo.

The Decision Framework: How to Actually Choose

You've now got a lay of the land for what's available. The question is how do you decide what's right for you?

Here's the framework I walk clients through.

First, prioritize your requirements. Start by listing your top three pain points in sales right now. Not theoretical nice-to-haves. Actual problems that are costing you deals or slowing you down. Then translate each pain point into a specific capability you need. For example, "reps waste time researching prospects" translates to "need enriched data with AI personalization." Then rank these capabilities by business impact. Which one, if solved, would move the needle most?

Your top three capabilities become your must-haves. Everything else is nice-to-have.

Second, evaluate technical requirements. Check your CRM compatibility - will this tool actually integrate? Look at API access - does your current CRM tier support the integrations you need? Consider data architecture - where will the data live and how will it flow? Think about existing integrations - will this play nice with tools you already have? And assess team technical skills - can your team actually implement and maintain this?

I've seen companies buy sophisticated AI platforms that required API expertise they didn't have. The platforms sat unused for six months until they finally hired someone who could implement them. Check your technical constraints before you buy.

Third, validate use cases. Don't just trust the demo. Ask for customer references in your industry at your revenue level. Run a pilot with a small subset of users before rolling out to everyone. Define specific success metrics before you start - how will you know if this is actually working? And set a timeline for the pilot - sixty to ninety days is usually enough to tell.

If a vendor won't let you pilot or won't give you relevant references, walk away. They're hiding something.

Fourth, calculate real ROI, not vendor ROI. The vendor's case study shows three hundred percent ROI. Great. What's YOUR expected ROI?

Take the annual cost of the tool including setup and training time. Then estimate the revenue impact - how much additional pipeline or closed revenue will this generate? Calculate the time savings - how many hours per rep will this save and what's the dollar value? Factor in adoption risk - what if only half your team actually uses it? And include hidden costs like ongoing training, integration maintenance, and data cleanup.

Vendor ROI assumes perfect execution. Your ROI needs to assume real-world challenges.

Here's a real example. Company evaluating Gong at sixteen thousand per user for eight users. Total cost is one hundred twenty-eight thousand per year. Vendor case study shows forty percent improvement in win rates. Sounds amazing.

But their actual analysis: Current win rate is twenty-five percent. Average deal size is seventy-five thousand dollars. Eight reps close an average of twelve deals per year each. That's ninety-six deals total at seventy-five thousand equals seven-point-two million in revenue.

If Gong improved win rate from twenty-five percent to thirty-five percent - that's ten percentage points which is a forty percent relative increase - they'd close one hundred thirty-four deals instead of ninety-six. That's thirty-eight more deals at seventy-five thousand equals two-point-eight-five million in additional revenue.

Two-point-eight-five million additional revenue for one hundred twenty-eight thousand investment equals twenty-two times ROI. Sounds great.

But then they factored in adoption risk. What if only seventy percent of reps fully adopt it? What if the win rate improvement is only twenty-five percent instead of forty percent? What if deal size doesn't stay at seventy-five thousand during a downturn?

Conservative estimate: twenty percent adoption discount, thirty percent results discount, ten percent deal size discount. Now the additional revenue is one-point-three-five million instead of two-point-eight-five million. That's still ten-point-five times ROI, which is great. But it's half of what the vendor case study implied.

Do this math for every tool. Don't just trust the vendor's numbers.

Budget Planning: The Tiered Approach to Building Your Stack

Most companies can't afford to buy everything at once. And honestly, you shouldn't. You need to build your stack in tiers based on what creates immediate value.

Here's how to think about budget allocation.

Tier 1 is the Foundation. This is your must-have infrastructure that everything else builds on. For most companies, this is CRM plus basic enrichment plus communication tools.

For small companies doing five to twenty million in revenue, your primary focus should be prospecting efficiency and basic automation. A good stack is Clay for enrichment at six hundred to one thousand per month plus HubSpot for CRM and marketing at six hundred to one thousand per month plus Otter for call transcription at one hundred per month. Total annual cost is sixteen thousand to twenty-five thousand. Expected ROI is three hundred to five hundred percent within twelve months just from time savings and better targeting.

For mid-market companies doing twenty to seventy-five million in revenue, you can add sophistication. Recommended stack is ZoomInfo or Clay for enrichment at two thousand to four thousand per month plus HubSpot or Salesforce for CRM at two thousand to three thousand per month plus Gong or Chorus for conversation intelligence at one thousand to two thousand per user per month. Total annual cost is sixty thousand to one hundred eight thousand. Expected ROI is two hundred to four hundred percent within twelve to eighteen months.

For enterprise companies over seventy-five million, you need integration and scale. Recommended stack is ZoomInfo for enrichment at five thousand to eight thousand per month plus Salesforce Enterprise for CRM at three thousand to five thousand per month plus Gong for conversation intelligence plus six sense for intent data plus People.ai for activity capture. Total annual cost is one hundred eighty thousand to three hundred sixty thousand. Expected ROI is one hundred fifty to three hundred percent within eighteen to twenty-four months.

Tier 2 is Intelligence and Insights. Once your foundation is solid, add tools that make your team smarter.

This includes conversation intelligence if you didn't already add it, predictive analytics for lead scoring and forecasting, intent data to know when accounts are in-market, and content intelligence if you have a large content library.

Add these six to twelve months after your foundation is working. Don't try to do everything at once.

Tier 3 is Advanced Capabilities. These are nice-to-haves that create incremental improvements once everything else is dialed in.

Think advanced analytics and attribution, sophisticated content personalization, AI-powered forecasting, and custom AI model development.

Most companies don't need Tier 3 until they're over one hundred million in revenue. Focus on getting Tier 1 and Tier 2 right first.

A Real Stack Selection Case Study

Let me show you how this works in practice.

Company was a forty-seven million dollar cybersecurity software company with a sales team of twelve reps. They came to me after spending six months evaluating AI tools and getting nowhere.

Their evaluation process had been a mess. They'd taken demos from twenty-three different platforms. Every demo looked good. Every vendor had impressive case studies. They had no framework for deciding.

We started with the current state assessment. They were on Salesforce Professional edition with decent data quality. Their sales process was documented but not consistently followed. They had no marketing automation. Their biggest pain points were prospecting taking too long, inconsistent messaging across reps, and no visibility into what was working in sales conversations.

We translated those pain points into requirements. They needed enriched data with personalization for prospecting efficiency. They needed conversation intelligence to understand what messaging worked. And they needed CRM and analytics tight enough to see the full funnel.

Based on their budget of seventy-five thousand per year and technical constraints with Salesforce Professional, we built a plan.

Phase 1 was immediate implementation of Clay for enrichment at eight hundred per month and Instantly for email deliverability at one hundred per month. This solved prospecting efficiency immediately. Cost was ten thousand eight hundred per year.

Phase 2 was three months later, after Clay was working, they upgraded Salesforce to Enterprise edition for better API access at an extra eighteen thousand per year. Then they added Gong for conversation intelligence at sixteen thousand per user for two sales managers as coaches. This solved messaging consistency. Cost was fifty thousand per year including the Salesforce upgrade.

Phase 3 was six months later, they added HubSpot Marketing Hub to connect with Salesforce for lead flow and nurturing. Cost was twelve thousand per year.

Total stack after twelve months was Clay plus Instantly plus Salesforce Enterprise plus Gong plus HubSpot Marketing. Total cost was seventy-two thousand eight hundred per year, just under their budget.

The results: Pipeline increased from one-point-eight million per quarter to four-point-one million per quarter. That's a two-point-three million increase. Win rates improved from eighteen percent to twenty-six percent because of better targeting and messaging consistency. Sales cycle decreased from one hundred nine days to eighty-four days because conversation intelligence helped reps handle objections better.

Additional revenue in year one was three-point-two million. Investment was seventy-two thousand eight hundred. ROI was forty-four to one.

But here's what mattered most: they built the stack systematically. Foundation first. Then intelligence. They didn't try to do everything at once. Each tool built on the previous one. By the time they added HubSpot Marketing, Clay and Gong were already working, so the integration was smooth.

That's how you do it right.

The Tools Don't Matter (But They Really Do)

Here's the paradox I need you to understand: The specific tools matter less than the system, but the tools you choose either enable or prevent you from building that system.

You could use different tools than what I've recommended here and still build an effective AI Sales Engine. Apollo instead of Clay. Chorus instead of Gong. Marketo instead of HubSpot. The exact vendor isn't what creates success.

What creates success is selecting tools that integrate properly, solve your specific pain points rather than theoretical nice-to-haves, fit your budget without overextending, match your team's technical capabilities, and most importantly, work together as a system rather than fighting each other for attention.

I've seen companies with expensive, sophisticated tools get mediocre results because the tools didn't integrate. And I've seen companies with modest, mid-tier tools crush it because everything worked together seamlessly.

The tool selection matters. But the system design matters more.

In Chapter 7, we're going to take the tools you select and show you exactly how to turn them into a data enrichment engine that gives your sales team everything they need to sell effectively. That's where the real magic happens - not in the tools themselves, but in how you orchestrate them.

But you can't build that engine with the wrong tools. So choose carefully. Choose systematically. And choose for integration, not just for features.

Executive Takeaways

For the CEO: Tool selection is a strategic decision, not a tactical one. The stack you build determines your sales team's effectiveness ceiling for the next two to three years. Don't delegate this entirely to your CTO or head of sales. You need to be involved in the decision framework even if not the specific vendor selection.

For the CRO: Resist the urge to buy everything that looks cool. Start with prospecting and enrichment - that's where you'll get the fastest ROI. Add conversation intelligence next if your average deal size justifies it. Save the fancy stuff for year two after your foundation is solid.

For the CMO: If you're not at the table for these decisions, you should be. Marketing and sales alignment starts with technology alignment. If sales is selecting tools without marketing input, you're going to end up with disconnected systems and misaligned data.

For the CFO: Budget for phased implementation, not big-bang deployment. Year one should focus on foundation at twenty to seventy-five thousand depending on company size. Year two adds intelligence at an additional twenty to fifty thousand. Budgeting it all upfront creates pressure to deploy everything at once, which usually fails.

For the CTO: Your role is translation and integration, not vendor selection. Help the business leaders understand technical constraints and integration requirements. Push back on tools that don't integrate. Demand pilot programs before enterprise commitments. And make sure someone owns the integration and maintenance, not just the initial implementation.

The companies that win with AI don't have the most tools. They have the right tools, implemented systematically, working together as a coherent system.

That's what we're building.

CHAPTER 7

Building Your Data Enrichment Engine

Here's what a typical day looks like for a sales rep without data enrichment.

Eight in the morning, they open their laptop and stare at a lead list. In Salesforce, it's a contact record with a name, an email address, and maybe a company name. In HubSpot, it's a row in a list view with the same three fields and a "last activity" date from four months ago. Everything else is blank. No title, no tech stack, no funding history, no recent news, no trigger events. Just a name and an email address - and somehow they're supposed to write a personalized, compelling message that makes this stranger want to take a meeting.

So they start researching. LinkedIn profile - open a tab. Company website - open another tab. Crunchbase for funding data - another tab. G2 or BuiltWith for tech stack - another tab. Google News for anything recent - another tab. By nine o'clock, they've got eight browser tabs open and they haven't written a single email. By eleven-thirty, if they're disciplined, they've sent maybe ten messages. The afternoon is follow-ups, a meeting or two, and prepping tomorrow's research batch.

Two and a half hours of detective work to send ten emails. That's not a sales team. That's a research department that occasionally sells something.

Now here's the same rep six weeks later, after a data enrichment engine is running.

Eight in the morning, they open their laptop and pull up their enriched Top 25 list. Every contact has a full profile - title, tenure, reporting structure, tech stack, recent funding, hiring signals, content engagement, trigger events. In HubSpot, the enrichment data populates custom properties that feed lead scoring and sequence enrollment automatically. In Salesforce, enriched fields drive assignment rules and priority views. The research is already done. By eight-fifteen, they're in full outreach mode - emails, calls, LinkedIn messages - all personalized using data they didn't have to hunt for. This continues until noon. The afternoon is follow-ups, meetings, and actual selling.

Same person. Same leads. Different data infrastructure.

But here's what matters - and it's not what you think. This isn't a productivity story. Sending more emails faster isn't the point. If it were, you could just buy a better sequencing tool and blast your whole database.

The point is that enriched data changes who your reps spend time on and why. Without enrichment, every lead looks the same - a name and an email - so reps either work the list alphabetically or cherry-pick based on gut instinct. The best prospects get the same treatment as the worst ones because nobody can tell them apart. Your highest-paid, most experienced closer is spending Tuesday morning researching a company that was never going to buy in the first place.

With enrichment, the system surfaces the twenty-five prospects who match your ICP, show active buying signals, and have a decision-maker in a new role with budget authority. Your closer isn't sending more emails - they're sending the right emails to the right people at the right time. The rep who used to book two meetings a week doesn't book eight because they got faster. They book eight because they stopped wasting time on prospects who were never going to convert and started focusing on the ones who were ready to have a conversation.

That's not a productivity improvement. That's a strategic advantage. And once you've seen a sales team operate with this kind of intelligence underneath them, going back to a blank

CRM record feels like sending someone into a negotiation blindfolded.

This chapter is about building the enrichment engine that makes it happen.

The Enrichment Gap Most CEOs Miss

Here's what most companies get wrong about AI sales: They think it's about the AI.

It's not. It's about the data.

AI doesn't create insights from thin air. It amplifies what the data tells it. Feed it incomplete data, you get incomplete results. Feed it rich, accurate, comprehensive data, you get results that look like magic.

But most companies treat data enrichment like it's optional - some nice-to-have thing that makes stuff "a little better."

That's like saying a foundation is optional when building a house because it just makes things "a little more stable."

Data enrichment isn't optional. It's the foundation of everything else you're about to build.

Let me show you why.

What Your ICP Blueprint Tells You vs. What Your Sales Team Actually Needs

Back in Chapter 5, you built your ICP Blueprint. It probably looks something like this:

Your target company has revenue between ten and fifty million dollars. They're in B2B SaaS. They've got fifty to five hundred employees. They're located in the United States. They use either Salesforce or HubSpot.

Your target buyer is a VP of Sales, Chief Revenue Officer, or Head of Sales. They're at the C-Suite or VP level. They sit in the sales or revenue department.

The pain points you solve? Manual prospecting eating their time. Inconsistent sales process across the team. Can't scale without losing quality. Sales cycles running too long.

That's a solid ICP. It's clear, it's specific, it's based on real data from your best customers.

But here's the uncomfortable truth: your sales team can't do anything with it.

Not because they're incompetent. Because there's a massive gap between "we know who to target" and "we have everything we need to actually reach them."

Your ICP says ten to fifty million in revenue. Your sales team needs the exact revenue figure, this year's growth rate, and whether they're profitable or burning through runway.

Your ICP says VP Sales or CRO. Your sales team needs the person's actual name, verified email that won't bounce, direct phone, LinkedIn profile, how long they've been in the role, where they came from, and what they post about on LinkedIn that might give you a conversation hook.

Your ICP says uses Salesforce or HubSpot. Your sales team needs the complete tech stack - every tool they use. They need to know about recent changes. Did they just add or remove something? What integrates with what? Are there obvious gaps your solution fills?

Your ICP says manual prospecting inefficiency. Your sales team needs actual evidence they're feeling this pain right now. Open sales roles because they're hiring. Recent funding because they have budget. New CRO because someone wants to make changes. Missing tech because they're using a CRM but doing personalization manually.

The ICP tells you where to aim. Enrichment tells you exactly what to say when you get there.

Companies that skip this step end up with sales reps writing emails like this: "Hi First Name, I noticed you work in software and thought you might be interested in our solution that helps companies be more efficient with their sales process\\..."

Generic. Forgettable. Deleted.

Companies that build proper enrichment engines write emails like this: "Sarah, saw you posted three sales roles last week - congrats on the growth. Most VPs I talk to at Series B companies scaling from eight to fifteen reps hit the same wall: how do you keep everyone consistent without drowning in coaching time? That's actually why I'm reaching out\\..."

Specific. Relevant. Gets responses.

The difference? Data enrichment.

The Five Levels of Data Enrichment: From Basic to Breakthrough

Not all data is created equal. There's a hierarchy - each level builds on the one before it.

Level one is firmographic data. This is the foundation. Company name, website, location, revenue, employee count, industry. It costs about a penny to a dime per record. You can get it from Clearbit, ZoomInfo, or Apollo. Accuracy runs around eighty-five to ninety-five percent.

This tells you whether the company is actually in your target market. It's like having someone's address before you send them mail. Without this, you can't do anything else.

With just firmographics, your outreach goes from "I help companies improve sales" to "I help thirty million dollar B2B SaaS companies improve sales." Slightly better. Still pretty generic.

Level two is technographic data. This is their tech stack - CRM, marketing automation, sales tools, analytics, how everything connects. It runs ten cents to fifty cents per record. You get it from BuiltWith, 6sense, or Datanyze. Accuracy is around seventy to eighty-five percent.

This tells you whether they have the foundation to use what you're selling. Are they already using competitors? What gaps exist in their stack? Tech stack is one of the strongest predictors of fit. Someone using Salesforce plus Outreach plus ZoomInfo is in a completely different universe than someone using spreadsheets and Gmail.

Now your outreach sounds like this: "Noticed you're using Salesforce and Outreach but handling personalization manually - that's usually where the bottleneck hits around fifteen reps\\..."

You're showing you understand their world.

Level three is people data. This is decision-maker identification. Names, titles, verified emails, phone numbers, LinkedIn profiles, reporting structure. It costs twenty-five cents to a dollar per contact. Sources include Apollo, ZoomInfo, Hunter, and RocketReach. Accuracy ranges from seventy to ninety percent.

This tells you who exactly you need to talk to and how to reach them. You don't sell to companies. You sell to people. All the firmographic and tech data in the world is worthless if you don't know who to contact.

Instead of emailing info at company dot com and hoping, you're now emailing sarah dot johnson at company dot com with ninety-two percent deliverability confidence. You're in the game.

Level four is intent and timing signals. This is where it gets interesting. Intent data tells you what they're researching. Timing signals tell you when to reach out. It costs fifty cents to two dollars per signal. You get it from 6sense, Bombora, LinkedIn, job boards, press releases. Accuracy varies widely, around sixty to eighty percent.

This tells you whether now is the right time. Are they actively looking? What's creating urgency?

The signals that matter: just raised funding, which means budget and expansion mode. New CRO hired, someone who wants to make impact. Posted five sales roles, they're scaling fast. Removed a competitor tool, window just opened. Downloaded a comparison guide, they're actively shopping.

Your outreach transforms: "Saw you just brought on five new reps based on LinkedIn - how are you planning to ramp them without losing consistency?"

You're showing awareness of their actual situation.

Level five is personalization data. This is the human layer. Individual interests, background, recent activities, shared connections, content they've created. It runs ten cents to fifty cents per person. You find it on LinkedIn, Twitter, blogs, podcasts, conference talks. Accuracy is high because it's observable public data.

This tells you what this specific person cares about and what will actually resonate. This separates good outreach from "holy shit, this person actually did their homework" outreach. It's the difference between eight percent response rates and eighteen percent.

The types of personalization that matter: previous company where you have a customer. School where they went to the same place or studied the same thing. Career path if they were promoted quickly or made interesting moves. Content they created like blog posts or podcast interviews. Interests like speaking at conferences or being active in specific communities. Recent changes like a new role or new company.

Your outreach now sounds like this: "I heard your interview on the Sales Enablement Podcast where you talked about scaling from ten to thirty reps without losing culture - that resonated because it's exactly what we help with. One of our customers at a similar company was in that exact spot\\..."

This is when prospects forward your email to colleagues saying "we need to talk to this person."

The Strategic Choice: Full Enrichment vs. Progressive Enrichment

Here's the reality: If you enrich every prospect to level five, you're looking at two to five dollars per record. For ten thousand prospects, that's twenty to fifty thousand dollars.

Most companies shouldn't spend that upfront.

The smarter play is progressive enrichment - enrich to different levels based on what you learn at each stage.

Stage one is initial screening. You apply levels one and two only - firmographics and technographics. It costs fifteen to sixty cents per record. You do this for all prospects in your initial target list, typically five to ten thousand companies. The purpose is figuring out who's actually a good fit. The output is a scored list separating winners from losers.

Stage two is priority enrichment. Now you add levels three and four - people data and timing signals. It costs seventy-five cents to two-fifty per record. You apply this to the top twenty percent from stage one, which might be a thousand to two thousand companies. The purpose is getting contact info and timing signals. The output is an actionable list ready for outreach.

Stage three is deep personalization. This is level five - the human layer. It costs ten to fifty cents per person. You do this for your top hundred to two hundred and fifty accounts, essentially your rotating Top 25 list over time. The purpose is maximum personalization for highest-value targets. The output is deeply researched accounts with custom hooks.

Here's the math on ten thousand prospects. You spend three thousand dollars enriching all ten thousand at levels one and two. Then three thousand dollars enriching two thousand priority prospects at levels three and four. Then seventy-five dollars enriching two hundred and fifty top accounts at level five. Total cost is six thousand and seventy-five dollars versus twenty to fifty thousand for full enrichment of everyone.

Progressive enrichment lets you invest where it matters without wasting money on prospects who won't pan out.

Why Enrichment Must Happen BEFORE CRM Entry

This is critical. Most companies get this backwards.

The broken way looks like this: A lead comes in with just a name and email from a form. The lead hits the CRM and gets assigned to a sales rep. The rep opens it and sees almost nothing. The rep then spends forty-five minutes researching - Google, LinkedIn, BuiltWith, news searches. The rep finally has enough context to write an email. The rep writes the email anyway, even though the research didn't really translate to good messaging. The prospect ghosts. The rep marks it "attempted contact" and moves on. The CRM fills with garbage data.

Time per lead? Sixty minutes. Leads processed per day? Six to eight. Response rate? Four percent. Rep satisfaction? "These leads suck." Data quality? Terrible.

The right way works differently. A target account gets identified from your ICP criteria. The account enters an enrichment workflow. The enrichment engine does its thing in about eight minutes, fully automated. It gathers company data, tech stack, decision-maker info, contact details, recent signals, and even generates AI-powered personalization. A fully enriched record lands in your CRM. The rep gets pinged that a high-fit prospect is ready. The rep opens it and sees everything - complete company profile, verified contact with deliverable email, recent signals like funding or hiring, an AI-generated opener ready to customize, suggested approach. The rep reviews it for two minutes, customizes the message for three minutes, sends it in one minute. Higher response rate because it's actually personalized. CRM data stays clean.

Time per lead? Six minutes of human time. Leads processed per day? Thirty-five to forty. Response rate? Fourteen percent. Rep satisfaction? "Finally, actual good leads." Data quality? Excellent.

The math is absurd. The old way with eight hours gives you eight leads, four percent response, point-three-two responses per day, one-point-six responses per week. The new way with the same eight hours gives you forty leads, fourteen percent response, five-point-six responses per day, twenty-eight responses per week. That's seventeen times more output. Same person. Same hours.

Your CRM should be a sales execution system, not a research project.

When Enrichment Changes Everything: A Client Case Study

Let me show you what happens when all five levels of enrichment come together in a real engagement.

One of our clients - a professional services firm - came to us with an ambitious growth mandate and a target list that would make most sales teams quit on the spot. Their Ideal Client Profile wasn't one audience. It was five.

The first ICP was Private Equity firms - partners and managing directors at mid-market PE shops actively deploying capital. The second was Finance executives - CFOs, controllers, and VP Finance at companies between fifty and five hundred million in revenue. The third was Technology leadership - CTOs, VPs of Engineering, and Chief Digital Officers at companies going through digital transformation. The fourth was Harvard Business School Alumni - not just anyone with an MBA, but HBS graduates in senior operating roles at portfolio companies and growth-stage firms. The fifth was McKinsey Partners - current and former McKinsey consultants who had moved into operating roles at the companies those PE firms were investing in.

Five ICPs. Five completely different data profiles. Five different sets of signals that indicate buying readiness.

If you tried to build this manually - researching each prospect one at a time across LinkedIn, Crunchbase, alumni databases, and consulting directories - you'd need a team of five SDRs working full-time for months. And by the time they finished the list, half the data would be stale.

We built the entire enrichment engine in Clay in under three weeks.

Here's what the system did. For Private Equity targets, it pulled firmographic data on fund size and investment thesis, identified recent deals and portfolio companies, found the specific partners responsible for each sector, verified their contact information through a four-provider waterfall, and scored them based on recent deal activity and fund deployment stage.

For Finance executives, it enriched company financials, identified whether they'd recently received funding or were in an active growth phase, pulled tech stack data to understand their current infrastructure, and generated personalized openers based on their company's financial trajectory.

For Technology leaders, it detected their current tech stack, flagged recent technology changes that signaled openness to new vendors, identified hiring patterns in engineering and product roles, and scored them on digital transformation readiness.

For the Harvard Business School Alumni program, we layered alumni database enrichment on top of standard firmographic and contact data. The system identified HBS graduates who were currently in C-suite or VP roles at companies that matched the firm's service profile, then enriched those records with company data, timing signals, and AI-generated personalization that referenced their shared alumni network.

For McKinsey Partners, the system identified former McKinsey consultants who had transitioned into operating roles - the people who understand strategic consulting, speak the language, and have both the authority and the analytical framework to evaluate complex service offerings.

Every record went through the same enrichment waterfall - firmographics, technographics, people data, timing signals, and AI-powered personalization. But the scoring criteria and personalization prompts were customized for each ICP. A PE partner who just closed a fund got a completely different message than a CTO who just ripped out their legacy CRM.

The enrichment engine produced a rotating Top 25 priority list for each ICP - one hundred and twenty-five high-priority, deeply personalized prospects refreshed every week.

But here's where it gets interesting. The client didn't just use this for cold email.

They used the enriched data to design a go-to-market strategy built around intimate, high-value events. Specifically, they organized C-suite dinners in three major cities - curated gatherings of twelve to fifteen senior executives per event, each one hand-selected based on their enrichment scores and ICP fit.

Think about what that requires. You're not blasting an invite to a bought list. You're personally inviting a CFO who just closed a Series C, a PE partner who just deployed a new fund, and a former McKinsey director who's now running operations at a portfolio company - and your invitation references something specific about each of their situations. That level of precision only happens when your data infrastructure is doing the heavy lifting.

The results were striking. Over one hundred attendees across the dinner series - nearly all of them C-suite or senior VP. Eight figures in qualified pipeline generated directly from those events and the follow-up sequences that came after. Seven figures in closed revenue within the first year.

Same firm. Same team. Same service offering they'd been selling for years. The difference was the data engine underneath their go-to-market motion.

This is what I mean when I say enrichment isn't a productivity tool - it's a strategic weapon. It didn't just help their sales team send more emails. It fundamentally changed what kind of go-to-market strategy was even possible.

How Enrichment Engines Actually Work: The Thirty-Thousand-Foot View

You don't need to understand every technical detail of how an enrichment engine is built to make the decision to build one.

But you do need to understand the core mechanics so you can evaluate vendors, ask the right questions, and hold your team accountable.

Here's how it works at a high level.

The Waterfall Principle

Traditional data providers work like a single pipe. You pay ZoomInfo or Apollo an annual contract, and they either have the data or they don't. Maybe they find the email forty to sixty percent of the time. But you paid for a hundred percent of the attempts.

Modern enrichment engines use what we call waterfall logic. Instead of relying on one provider, you stack multiple providers in sequence - cheapest first, most expensive last. The system tries the first provider. If it finds the data, you're done. If it doesn't, it automatically tries the next one. And the next. And the next.

The result is eighty-five to ninety-five percent enrichment success rates instead of forty to sixty percent. And you only pay for successful enrichments at each tier.

On a typical thousand-prospect enrichment run, the waterfall approach delivers four times more successful enrichments at roughly one-fifth the cost of a traditional single-provider contract. That's not a marginal improvement - it's a fundamentally different economics model.

Progressive Enrichment: Spend Where It Matters

The second key principle is progressive enrichment - don't spend the same amount enriching every prospect.

At the top of the funnel, you enrich broadly but cheaply. Firmographic and technographic data for your entire target list at fifteen to sixty cents per record. This tells you who's actually in your ICP and who isn't.

In the middle, you enrich deeply but selectively. People data, contact verification, and timing signals - but only for the top twenty percent that passed your initial screening. This costs seventy-five cents to two-fifty per record, but you're only spending it on prospects who've already demonstrated fit.

At the bottom, you enrich intensively but narrowly. Deep personalization, AI-generated messaging, buying committee mapping - but only for your Top 25 to Top 100 priority accounts. These are the prospects that justify ten to fifty cents per person in personalization investment because the deals are worth it.

The math works beautifully. Ten thousand prospects enriched progressively costs six to eight thousand dollars. The same ten thousand enriched to full depth across the board would cost twenty to fifty thousand. You save eighty percent and actually get better results because your highest-value prospects get more attention, not less.

AI-Powered Personalization at Scale

The third principle that makes modern enrichment engines transformative is AI integration. Platforms like Clay have native GPT integration, which means you can feed all your enriched data into an AI prompt during the enrichment process itself.

The system takes everything it knows about a prospect - their company's funding history, their tech stack, their hiring signals, the decision-maker's background and recent career moves - and generates a personalized opening line, email subject line, or call preparation brief.

Your sales reps don't just get a spreadsheet of data. They get actionable messaging ready to customize and send.

The quality isn't perfect - you'll want a human reviewing the AI outputs, especially for your highest-priority accounts. But it's good enough that a rep who used to spend forty-five minutes researching and writing a single email can now review, customize, and send a deeply personalized message in five to six minutes.

Scoring and Prioritization: The Top 25

The final piece is systematic scoring. Every enriched prospect gets a composite score based on firmographic fit, technographic fit, timing signals, data completeness, and personalization quality. Scores are weighted based on what matters most for your specific business.

The output is a ranked priority list - your Top 25 - that refreshes weekly as new signals come in. A company that scored moderately last week might jump to the top of the list this week because they just posted six sales roles or announced a new funding round.

The discipline of focusing your sales team on twenty-five deeply researched, perfectly timed, fully personalized prospects per week - instead of spraying hundreds of generic emails - is what produces twelve to eighteen percent response rates instead of four to five percent.

The Platform Landscape

For most companies in the five to one hundred million revenue range, Clay is the right platform for building enrichment engines. It offers the waterfall flexibility, AI integration, and usage-based pricing that mid-market companies need.

Enterprise companies over one hundred million may prefer ZoomInfo for its managed service and guaranteed SLAs. Smaller companies under twenty million might start with Apollo's all-in-one simplicity. And many companies end up with a hybrid approach - Clay for enrichment and personalization, ZoomInfo or Apollo for deep contact data on priority accounts, and a separate email tool for deliverability.

The platform matters less than the principles. Waterfall enrichment, progressive investment, AI-powered personalization, and systematic scoring work regardless of which tool you use.

A Note on What Comes Next

What I've just described is the strategic framework - the what and the why of data enrichment engines. The actual implementation involves configuring waterfall sequences provider by provider, building scoring formulas with weighted criteria, setting up automated refresh schedules, tuning AI prompts through iterative testing, and connecting everything to your CRM and email infrastructure.

It's detailed, technical work. Some of you will want to build it yourselves - and you absolutely can. Others will want someone to build it for them, which is what our team does every day.

Either way, here are two resources to help you move forward.

For a complete, step-by-step walkthrough of how to build an enrichment engine from scratch - including the exact Clay configurations, provider recommendations, scoring formulas, and AI prompts we use with clients - visit foleymedia.ai/clay. You'll find a full platform demo that walks through everything at the implementation level.

And if you'd rather skip the learning curve and have our team build it for you, book a call at FoleyMedia.ai. We'll handle the assessments, the setup, the workflow design, and the ongoing optimization. We've done this dozens of times across every industry and company stage you can imagine.

The goal is getting an enrichment engine running - not winning a merit badge for building it yourself.

From Enrichment to Activation: Making Data Usable for Your Sales Team

You built a beautiful enrichment engine. Your Clay table is full of perfectly scored, deeply personalized prospects. Now what?

This is where most companies fail. They have great data but don't format it for easy sales team consumption. The sales team stares at a spreadsheet with fifty columns and doesn't know what to do with it.

The handoff from enrichment to activation is absolutely critical.

You need three different export formats depending on how your team works.

Format one is the daily Top 25 hit sheet. This is what your sales team sees every single day. It lives in a Google Sheet with one tab called "Today's Top 25."

The structure includes rank number so they know priority order. Company logo for visual recognition. Company name and contact name with their title. Email address ready to copy and paste. ICP fit score showing confidence level. A fire emoji with the key signal explaining why now. The personalization hook ready to use or customize. Recommended channel showing email versus call versus LinkedIn. And a status checkbox for tracking.

The sales rep workflow is dead simple. Eight AM, they open the sheet. Eight-fifteen, they pick five to eight accounts to focus on today. Eight-thirty, they start outreach with no research needed because it's all done. Throughout the day, they check boxes as they go. Five PM, they update status on any responses or meetings.

Format two is the account deep-dive one-pager. Use this for high-value accounts worth one hundred thousand plus or strategic accounts that would make great case studies.

The one-pager includes the account name with their ICP fit score like one forty-five out of one sixty showing perfect fit. A company overview section with their industry, size, funding stage, and headquarters. A primary contact section with name, title, email, LinkedIn, plus context like "Joined sixty-two days ago" and "Previously VP Sales at competitor company." A technology stack section showing what they currently use, what they're missing that you fill, and any recent changes. A timing signals section highlighting recent funding, leadership changes, team expansion, and why now is the perfect time. Personalization hooks section with four to five specific conversation starters. A recommended outreach approach section with email copy, follow-up strategy, and suggested sequence. And a likely objections section with how to respond to each one.

When you create deep-dives, do it for your top ten accounts by priority score, accounts worth over one hundred thousand dollars in potential annual recurring revenue, strategic accounts that would be great references or case studies, and complex deals with multiple stakeholders in the buying committee.

Format three is the CRM-ready import file. This is for getting enriched data into your CRM system, which we'll cover in detail in Chapter 8.

The CSV structure includes all standard CRM fields like first name, last name, email, title, phone, company name, domain, revenue, employees, industry. Plus custom fields that you create including ICP underscore Fit underscore Score, Technology underscore Stack, Timing underscore Signals, Personalization underscore Hook, Enrichment underscore Date, and Data underscore Source.

What's critical for CRM import: All required fields must be populated. Data must be formatted correctly - dates as dates, numbers as numbers, text as text. Custom fields must be created in your CRM before importing. Duplicates must be removed. And test the import with ten records before doing the full batch.

Before you hand enriched data to your sales team, run through this pre-activation checklist.

For data accuracy: Random sample of twenty-five records manually verified. Error rate under ten percent confirmed. All Top 25 records individually verified because you have zero tolerance for errors in highest priority.

For enrichment completeness: Ninety-five percent or higher have verified email addresses. Ninety percent or higher have decision-maker identified. Eighty percent or higher have firmographic data complete. Seventy percent or higher have at least one timing signal. One hundred percent of Top 25 have personalization hooks that were reviewed by a human.

For scoring validation: Top 25 list reviewed by sales leadership. Confirmed these actually match ICP intent and aren't just algorithm artifacts. Removed any obvious mismatches or edge cases. Scoring criteria documented and agreed upon by the team.

For sales team readiness: Team completed the thirty-minute training on how to read enriched data. Playbook created for using personalization hooks effectively. Process documented for updating status and results. Manager assigned to monitor adoption and quality. For the first week, manager reviews every outreach message before sending to ensure quality.

For technical setup: Export format tested and it imports cleanly to CRM or loads properly in sheets. All custom fields mapped correctly between Clay and CRM. Automation triggers tested if you're using Zapier or Make. Backup process documented in case something breaks.

Only activate once ALL boxes are checked.

I've seen companies rush this step and waste months of work because their sales team didn't know how to use the data or because data quality was poor. One client activated with a twenty-five percent error rate because they skipped the quality audit. Their sales team reached out to prospects with wrong names, wrong companies, and incorrect personalization. Response rate was two percent and prospects were annoyed. We paused everything, fixed the enrichment workflow, re-enriched the data, verified quality properly, retrained the team, and relaunched. Response rate jumped to sixteen percent.

The lesson is clear: speed to activation is completely useless if the activation is broken.

Measuring Enrichment ROI: Proving the Value to Your CFO

Your CFO doesn't care about enrichment workflows or Clay tables or waterfall strategies. They care about return on investment.

Here's how to measure and communicate the business impact of your enrichment engine.

In week one, the immediate ROI is time savings. Before enrichment, your sales rep researches for two to three hours per day. They process five to eight prospects per day. Time per prospect is forty-five to ninety minutes. After enrichment, sales reps review enriched data for fifteen minutes per day. They process thirty to forty prospects per day. Time per prospect is six to eight minutes.

The time savings calculation works like this. Two and a half hours saved per day per rep. Multiply by five days equals twelve and a half hours per week per rep. Multiply by fifty dollars per hour loaded cost equals six hundred twenty-five dollars per week per rep. Multiply by five reps equals three thousand one hundred twenty-five dollars per week. Multiply by fifty-two weeks equals one hundred sixty-two thousand five hundred dollars per year in time savings.

Your enrichment cost is three to six thousand dollars per year. ROI from time savings alone is twenty-seven to fifty-four times your investment.

In month one, the next ROI is response rate improvement. Before enrichment, response rate is five percent from generic outreach. One hundred emails sent equals five responses. After enrichment, response rate is fourteen percent from personalized outreach. One hundred emails sent equals fourteen responses.

The impact is two-point-eight times more responses from the same effort. Or you can reach your original goal with sixty-five percent less outreach, which saves time and improves deliverability.

The meeting booking calculation shows responses that convert to meetings at about twenty percent. Before enrichment, five responses times twenty percent equals one meeting per one hundred emails. After enrichment, fourteen responses times twenty percent equals two-point-eight meetings per one hundred emails. That's two-point-eight times more meetings from the same outreach volume.

In quarter one, the ultimate ROI is pipeline generation. Before enrichment, per sales rep per quarter: outreach volume is five hundred emails, response rate is five percent, meetings are five per rep, opportunities created are one-point-five per rep at thirty percent of meetings, average opportunity value is fifty thousand dollars, pipeline per rep is seventy-five thousand dollars.

After enrichment per sales rep per quarter: outreach volume is fifteen hundred emails because they're three times faster, response rate is fourteen percent, meetings are forty-two per rep, opportunities created are twelve-point-six per rep at thirty percent of meetings, average opportunity value is still fifty thousand dollars, pipeline per rep is six hundred thirty thousand dollars.

Pipeline increase is five hundred fifty-five thousand dollars per rep per quarter. For a five-person sales team, that's two-point-seven-seven million dollars additional pipeline per quarter.

Enrichment cost for the quarter is fifteen hundred dollars. ROI is one thousand eight hundred fifty times your investment.

Using a conservative twenty percent close rate, that five hundred fifty-five thousand in additional pipeline turns into one hundred eleven thousand in additional revenue per rep per year. With enrichment investment of six thousand dollars per year, your revenue ROI is ninety-two times.

By the end of year one, you can measure complete business impact. Revenue impact shows additional pipeline generated of eleven million dollars across the team. Additional closed revenue at twenty percent close rate is two-point-two million dollars. Enrichment investment was six thousand dollars. Revenue ROI is three hundred sixty-seven times.

Efficiency impact shows sales rep productivity increased one hundred eighty-five percent. Cost per meeting decreased from eight hundred fifty dollars to one hundred forty-eight dollars. Time to first meeting decreased from twelve days to three-point-two days. Sales cycle length decreased twenty-three percent because better targeting leads to faster closes.

Strategic impact includes sales team morale up forty-five percent because they're working with better leads. CRM data quality improved from forty percent accuracy to ninety percent. Forecast accuracy improved sixty-seven percent because better data enables better prediction. Competitive win rate increased thirty-four percent because personalization differentiates you.

When you present enrichment ROI to your board or executive team, here's the slide that matters.

Title it "Data Enrichment Engine: Year 1 Results." Show investment of six thousand dollars. Then show return broken into three categories.

Revenue impact: Additional pipeline eleven million dollars. Additional revenue two-point-two million dollars. Revenue ROI three sixty-seven to one.

Efficiency gains: Sales productivity up one eighty-five percent. Cost per meeting down eighty-three percent from eight-fifty to one forty-eight. Time to first meeting down seventy-three percent from twelve days to three-point-two days. Sales cycle down twenty-three percent.

Quality improvements: Response rate up one eighty percent from five percent to fourteen percent. CRM data accuracy up one twenty-five percent from forty to ninety. Forecast accuracy up sixty-seven percent. Team satisfaction up forty-five percent.

The bottom line: Every dollar invested in enrichment returned three hundred sixty-seven dollars in revenue. Same sales team, same hours worked, fundamentally better results.

Alternative Enrichment Approaches for Different Company Stages

Clay is powerful, but it's not right for every situation. Let me show you three alternative approaches and when to use them.

Alternative one is the ZoomInfo-first approach for enterprise companies. This makes sense when you're over one hundred million in revenue with a sales team of fifty-plus people. You have fifty thousand dollars or more per year in committed budget. You need guaranteed accuracy SLAs, dedicated support, and compliance infrastructure for regulated industries. Your use case is enterprise selling, highly regulated industries, or large account-based programs.

The ZoomInfo-first workflow uses ZoomInfo as your primary database and enrichment source with native CRM integration to Salesforce or HubSpot. Then you use Clay as a supplement for custom AI personalization since ZoomInfo doesn't have GPT integration, for enrichments ZoomInfo doesn't offer like tech change signals and hiring data, and for cost optimization on non-critical enrichments where you can use cheaper providers when quality differences don't matter much.

The cost structure is ZoomInfo at fifty thousand per year on a committed contract plus Clay at two thousand per year for supplemental personalization and custom signals. Total is fifty-two thousand per year.

Choose this when you need someone to call when things break - enterprise support matters. When compliance is critical in healthcare, finance, or government sales. When data accuracy requirements are extreme at over ninety-five percent required. When budget is available and already approved by finance.

The ROI consideration is that ZoomInfo's higher cost is worth it IF your deals are one hundred thousand dollars plus where the personalization improvement pays for the cost. If your team is large enough that implementation simplicity matters at scale. And if your compliance risk is high enough that you need the insurance of proper vendor management.

Alternative two is the Apollo-based approach for small teams. This makes sense when you're five to thirty million in revenue with a sales team of two to ten people. You need all-in-one simplicity with prospecting, enrichment, and outreach in one platform. Your budget is five to fifteen thousand dollars per year total. Your use case is straightforward B2B outbound with no complex workflows needed.

The Apollo all-in-one workflow uses Apollo for everything - prospecting, enrichment, AND email sequences all in the same platform. Setup is simple with no integration complexity. You only add Clay if you need better email deliverability through waterfall enrichment that Apollo doesn't offer, or if you need advanced AI personalization beyond Apollo's basic templates.

The cost structure is Apollo at ten thousand per year for prospecting plus enrichment plus sequences for the whole team. Clay is zero dollars unless you add it later for specific gaps. Total is ten thousand per year.

The pros versus Clay are simpler setup with one platform and no integrations, faster to launch at one week versus two to three weeks with Clay, lower learning curve so SDRs are productive immediately, and all-in-one UI where prospecting, enrichment, and outreach happen in the same place.

The cons versus Clay are less enrichment flexibility with a single data source, no waterfall strategy so you get seventy percent enrichment rate versus ninety percent with Clay, limited AI personalization with basic templates versus GPT-powered with Clay, and less cost-efficient at scale because of fixed pricing versus usage-based.

Choose Apollo if simplicity is more important than optimization for your stage. Choose Clay if data quality and personalization are critical to your business model.

A real example shows a twelve million dollar SaaS company with four SDRs started with Apollo for simple setup and fast launch. After six months of proving the model worked, they upgraded to Clay plus Instantly for better enrichment quality and AI personalization. Response rates improved from nine percent with Apollo to fifteen percent with Clay. The progression was correct - start simple, prove it works, then optimize.

Alternative three is the hybrid approach, which I recommend for most companies between ten and one hundred million in revenue. This is the best-of-both-worlds strategy.

The hybrid stack works in four phases. Phase one is discovery using Clay. Build your initial target account list of one to five thousand companies. Enrich with Clay to level one and two only - firmographics and technographics. Score and filter to identify the highest-fit accounts. Cost is one to three thousand dollars one-time.

Phase two is deep enrichment using ZoomInfo or Apollo. Take the top two hundred fifty accounts from Clay analysis. Do deep enrichment with ZoomInfo or Apollo to get multiple contacts per account. Get the full buying committee of three to five people. Add phone numbers for priority accounts. Cost is five to ten thousand dollars per year for limited licenses.

Phase three is personalization using Clay again. Push all enriched data back to Clay. Run AI personalization workflows to generate customized talking points, email openers, and call scripts. Cost is five hundred to one thousand dollars per year.

Phase four is activation using email tool plus CRM. Export to HubSpot or Salesforce CRM for pipeline management. Sync to Instantly or Smartlead for email sequences and deliverability. Launch multi-channel campaigns. Cost is three to eight thousand dollars per year.

Total hybrid cost is nine thousand five hundred to twenty-two thousand per year. Compare to ZoomInfo enterprise-only at fifty thousand plus per year. You're saving fifty-five to eighty percent with better flexibility.

Why the hybrid approach works is you get Clay's enrichment flexibility for initial screening, ZoomInfo's data accuracy for your final priority prospects, Clay's AI personalization for messaging, purpose-built email tools for deliverability, and cost optimization where you pay premium prices only where it matters most.

What you avoid is Clay's learning curve for your entire team since only RevOps needs to know Clay. ZoomInfo's cost for low-priority prospects that aren't worth the premium pricing. And single-platform lock-in since you use the best tool for each specific job.

Implementation timeline is week one through two, set up Clay enrichment workflows. Week three, enrich one thousand companies, score them, identify top two fifty. Week four, purchase ZoomInfo credits for top two fifty and do deep enrichment. Week five through six, run AI personalization in Clay and export to CRM. Week seven through eight, set up email sequences and launch campaigns. Total is eight weeks from start to activation.

A real client example shows a forty-five million dollar B2B SaaS company with a fifteen-person sales team. Their hybrid stack was Clay for initial enrichment of five thousand prospects to identify five hundred high-fit accounts at twenty-five hundred dollars. ZoomInfo with five licenses for deep enrichment of top five hundred accounts including buying committees at twelve thousand per year. Clay for AI personalization for all five hundred accounts at eight hundred per year. HubSpot for CRM and pipeline management at fifteen thousand per year for the whole team. Instantly for cold email sequences at twelve hundred per year. Total was thirty-one thousand five hundred per year.

Results in year one were pipeline generated of eighteen-point-five million dollars. Closed revenue of four-point-one million. ROI of one thirty to one comparing revenue to stack cost. Sales team productivity increased two hundred twenty percent.

The CEO's quote captures it: "We almost bought the full ZoomInfo enterprise package for sixty thousand. This hybrid approach saved us thirty thousand and actually works better because we're not enriching garbage data - we're enriching already-qualified accounts."

Executive Takeaways: What Each Leader Needs to Know

For the CEO, understand that data enrichment isn't a sales tool - it's a competitive advantage. Companies with systematic enrichment engines are outperforming competitors three to five times on the same budgets and headcount.

The key decisions you need to make are budget allocation of five hundred to three thousand per month for enrichment infrastructure, team ownership of who owns enrichment workflows whether it's RevOps, Sales Ops, or Marketing Ops, and timeline of three to four weeks to build, two to three months to prove ROI, and twelve months to reach full maturity.

Expected ROI is fifty to one hundred times in year one, and that's a conservative estimate based on time savings plus effectiveness improvement.

For the CRO, understand that your sales team's productivity ceiling is determined by data quality, not by training or motivation. The best closer in the world can't close bad data.

What changes with proper enrichment: Sales reps spend ninety-five percent of their time selling versus forty percent before. Response rates improve two to three times because personalization actually works. Pipeline quality improves two times or more because better targeting means better fit. Sales team satisfaction increases dramatically because they finally have leads that don't suck.

Your KPIs will improve across the board. Time to first meeting decreases sixty to eighty percent. Cost per meeting decreases seventy to eighty-five percent. Sales cycle length decreases twenty to thirty percent. Win rate increases twenty-five to forty percent.

Implementation risk is low. You can test with twenty-five to fifty prospects for under one hundred dollars before committing to anything larger.

For the CMO, enrichment is the bridge between marketing and sales. It transforms marketing-qualified leads, which are often just form fills, into sales-qualified prospects with deep research ready for personalized outreach.

Marketing-sales alignment improves when both teams work from the same enriched data as a single source of truth. Lead scoring incorporates enrichment signals like funding, hiring, and tech changes instead of just form fills and page views. Marketing campaigns can leverage intent data for perfect timing. Content can be personalized at scale because AI uses enrichment data to customize.

Your programs get better across the board. ABM campaigns become truly account-based with deep account intelligence. Lead nurturing becomes genuinely relevant because timing signals drive cadence. Content recommendations become accurate because tech stack indicates needs. Attribution becomes clearer because you can see what signals drive conversion.

For the CFO, here's the business case in three numbers. Investment is six to thirty-six thousand dollars per year depending on volume and approach. Return is three hundred thousand to two million dollars in additional revenue in year one, varying by deal size and team size. ROI is fifty to one hundred times conservatively, potentially two hundred to three hundred times in best cases.

The cost structure breaks down as year one at six to thirty-six thousand for implementation plus first year operation. Year two and beyond at three to eighteen thousand per year for maintenance plus refresh. Payback period is typically one to three months.

Risk mitigation includes starting with a pilot of twenty-five to fifty prospects costing one hundred to five hundred dollars for testing. Using usage-based pricing where you pay for what works instead of fixed contracts. Having month-to-month commitment with no long-term lock-in with the Clay approach.

Compare to alternatives: hiring an additional SDR costs sixty to eighty thousand per year plus ramp time. Enterprise data platform costs fifty to one hundred thousand per year committed. Manual research costs thirty to fifty thousand per year in wasted sales time. The enrichment engine approach saves eighty to ninety percent with better results.

For the CTO or RevOps leader, understand that implementation is medium complexity. This is a Zapier-level project, not a custom development effort.

Technical requirements include a Clay account with no coding required but spreadsheet-level complexity needed. API connections to data providers that are built into Clay. CRM integration via Zapier, Make, or native depending on which CRM you use. Email tool integration using webhooks, which is straightforward.

Implementation timeline is week one through two to build enrichment workflows in Clay. Week three to test on twenty-five to one hundred prospects and verify quality. Week four to integrate with CRM. Week five through six to connect email automation. Week seven through eight to train team and launch. Total is six to eight weeks with existing resources and no consultants needed.

Ongoing maintenance is three to five hours per week once running smoothly. Weekly Top 25 refresh is automated with just a spot-check needed. Monthly cost optimization review to ensure you're not wasting money. Quarterly scoring model refinement based on what's actually converting.

Technical debt is low because you're using modern SaaS tools with standard integrations and no custom code. Skillset required is someone comfortable with spreadsheets and logical thinking. You don't need a developer.

You've now learned exactly how to build a systematic data enrichment engine that can transform your sales team's effectiveness by three to five times.

The key insight is this: AI-powered sales doesn't start with AI tools. It starts with AI-ready data. Your enrichment engine is the foundation that everything else gets built on top of.

In Chapter 8, we're going to take this enriched data and show you exactly how to activate it. We'll cover the progression from spreadsheet testing to CRM implementation to automated email sequences to full multi-channel campaigns.

But you can't skip this chapter. Without proper enrichment, everything in Chapter 8 will fail. The activation is only as good as the data feeding it.

As one of my clients put it after implementing their enrichment engine: "We spent two years trying to fix our sales process. Turns out the problem wasn't the process - it was that we were asking our team to sell with one hand tied behind their back. Enrichment untied that hand."

Now let's put that enriched data to work.

CHAPTER 8

CRM Integration & Activation Systems

The number one reason AI sales implementations fail isn't bad technology. It isn't bad data. It's that someone tried to change the entire sales team's workflow on a Tuesday and expected them to hit quota on Wednesday.

You've seen this play out. The enrichment engine is built. The data is flowing. Leadership is excited. So someone makes the call: "Let's get everything into the CRM this week, set up the sequences, turn on the automation, and go live by Monday." The sales team walks in on Monday morning to a system they've never used, workflows they've never seen, and a quota that hasn't changed. By Wednesday, half of them have gone back to their personal spreadsheets. By Friday, your best rep is in your office telling you this isn't working.

It's not that they're resistant to change. It's that you dropped a new system on them without giving them a reason to trust it, a chance to learn it, or any proof that it actually works better than what they were doing before. You asked them to take a leap of faith in the middle of a quarter where their paycheck depends on closing deals, not learning software.

The second mistake is just as common and just as deadly: taking your old process and running it on new tools. Companies spend months implementing Salesforce or HubSpot, migrate all their data, configure all their automations - and then use the new platform to do exactly what they were doing before, just with a fancier interface. That's not transformation. That's a $50,000 skin on the same broken process.

New tools require new processes. And new processes require proof before people will adopt them.

This chapter is about getting both right - phasing the transition so your team sees results before they're asked to change, and redesigning the workflow so the tools actually unlock something that wasn't possible before. The companies that nail this don't just get adoption. They get a sales team that can't imagine going back.

The Spreadsheet-to-CRM Philosophy

Here's what most consultants won't tell you: spreadsheets are often the right answer, at least initially. Not because they're better than CRMs. Because they're simpler, faster to implement, and easier to change when you're still figuring out what works.

The companies that succeed with AI sales systems don't start with perfect technology. They start with perfect processes. Then they upgrade the technology to scale what's already working.

I learned this lesson the hard way with a client that insisted on implementing Salesforce, HubSpot sequences, and Outreach automation simultaneously. Six months and tens of thousands of dollars later, their results were worse than when they started. The team was overwhelmed by complexity. The management team couldn't figure out what was working. The data was scattered across three platforms with inconsistent field mapping.

We had to start over with spreadsheets.

Within four weeks using simple Google Sheets, they had better results than six months of enterprise software had delivered. Why? Because they could focus on methodology instead of technology. They could iterate messaging instead of troubleshooting integrations. They could understand cause-and-effect instead of hoping the "AI" would figure it out for them.

This is the principle that should guide every technology decision in your AI Sales Engine: prove the process first, then invest in the infrastructure to scale it. Forrester Research reports that nearly half of all CRM projects ultimately fail, and the leading cause isn't technical issues or budget constraints - it's poor user adoption. When you prove the process before you implement the technology, adoption isn't a problem because the team already trusts the methodology.

Why This Sequence Matters

The spreadsheet-first philosophy isn't about avoiding technology. It's about implementing technology in the right sequence for four strategic reasons.

First, risk management. Zero implementation cost, no long-term contracts, no integration complexity. If your targeting assumptions are wrong or your messaging doesn't resonate, you've invested time but not capital. You can pivot in hours rather than weeks.

Second, faster validation. No setup time, no training lag, no approval processes. You can test a completely new ICP hypothesis, messaging approach, or outreach cadence today and have meaningful data by Friday. That speed of learning is impossible once you've locked into CRM workflows and automation sequences that took weeks to configure.

Third, natural team buy-in. Everyone knows how to use spreadsheets. The resistance your team feels isn't about learning software - it's about changing how they work. When the process is proven in a familiar tool and they can see the results with their own eyes, technology adoption becomes a welcome upgrade rather than an unwanted mandate. They're asking for the CRM instead of being forced into it.

Fourth, clear requirements. Four weeks of spreadsheet-based operations will show you exactly what functionality you need from a CRM. You're buying tools to solve known problems instead of hoping tools will solve unknown problems. Every custom field, every automation trigger, every reporting dashboard is driven by actual needs rather than theoretical use cases that sounded good in a vendor demo.

When to Skip Straight to CRM

Not every company needs the spreadsheet phase. If you already have a CRM in production with established data and workflows, it makes no sense to regress to spreadsheets. If your team exceeds ten salespeople, spreadsheet coordination becomes chaotic quickly. If your sales cycle is long and multi-stakeholder, you need the tracking capabilities from day one. And if you have dedicated RevOps resources ready to configure and maintain the system, the spreadsheet phase just slows them down.

The honest assessment is about where your organization actually is, not where you want it to be. Growth-stage companies without dedicated operations resources almost always benefit from proving their methodology in spreadsheets first. Established companies with existing infrastructure should integrate enrichment into what they already have.

Choosing the Right CRM Platform

CRM selection is one of the most consequential technology decisions in your AI Sales Engine build. The wrong choice creates friction that compounds daily. The right choice creates leverage that compounds equally.

The market has consolidated around three tiers, each optimized for different organizational profiles. Understanding which tier fits your situation prevents the expensive mistake of buying capability you'll never use or, worse, outgrowing your platform within a year.

The Growth-Stage Platform: HubSpot

HubSpot has become the default choice for companies in the five to fifty million dollar revenue range, and for good reason. It balances sophisticated AI sales features with genuine usability. Sales teams become productive in days rather than weeks, which matters enormously when adoption determines success.

The strategic rationale for HubSpot centers on marketing-sales alignment. HubSpot was built as an integrated platform, so the handoff between marketing lead generation and sales follow-up is native rather than bolted on. For companies where marketing and sales teams need shared visibility into the prospect journey, this integration eliminates the data silos that plague organizations using separate marketing automation and CRM platforms.

Clay integration is seamless through HubSpot's app marketplace with native, bi-directional sync. Your enrichment data flows directly into CRM records where reps can use it during outreach. The custom property system accommodates ICP scores, personalization hooks, timing signals, and every other enrichment field from your Clay workflows without requiring developer resources.

The AI features that matter for your sales engine include native lead scoring that incorporates enrichment data, email sequencing with AI-optimized send times, conversation intelligence with call recording and analysis, workflow automation triggered by prospect behavior, and predictive deal forecasting. These aren't theoretical features - they're the operational backbone of the systematic approach this book teaches.

The limitation is enterprise complexity. Once your organization exceeds fifty salespeople, requires complex territory management, or needs deep customization beyond what the platform offers natively, you'll feel the ceiling.

The Enterprise Platform: Salesforce

Salesforce is the choice for organizations that need maximum customization, have dedicated RevOps resources, and operate at enterprise scale. The platform can do virtually anything, which is both its greatest strength and its primary risk.

The strategic rationale for Salesforce centers on extensibility. Einstein AI features are best-in-class for lead scoring, opportunity prediction, and email optimization. The ecosystem of third-party integrations is unmatched. Custom development capabilities enable workflows that no other platform can replicate. For organizations with complex sales processes involving multiple business units, international operations, or regulatory requirements, Salesforce is often the only platform that can accommodate every need.

Clay integration requires API setup or middleware like Zapier, which adds a layer of complexity but also adds flexibility. Custom objects can accommodate any enrichment data structure, and Process Builder enables sophisticated automation workflows that respond to enrichment signals in real time.

The risk is implementation complexity. Gartner's 2024 research found that seventy percent of sellers reported being overwhelmed by the number of technologies required to do their work. Salesforce, configured poorly, contributes significantly to that overwhelm. The platform requires dedicated technical resources for configuration, maintenance, and optimization. Without them, you end up with an expensive system that the sales team resents and underutilizes.

The honest assessment: if you don't have at least one person whose primary responsibility is CRM administration, Salesforce will likely become a liability rather than an asset. The power is real, but so is the operational burden.

The Agency and SMB Platform: Go High Level

Go High Level occupies a unique position for marketing agencies managing multiple clients and small businesses wanting all-in-one functionality at aggressive pricing. Unlimited users across multiple business units for a fraction of enterprise platform costs makes the economics compelling for specific use cases.

The strategic rationale centers on consolidation and cost efficiency. Rather than paying separately for CRM, email automation, SMS, booking, and reputation management, GHL bundles everything into a single platform. For agencies that need to replicate their sales methodology across multiple client accounts, the multi-tenant architecture is purpose-built.

Clay integration works through Zapier with webhook triggers for automation, and API access enables custom integrations for teams with technical resources.

The trade-off is sophistication. GHL's individual features don't match the depth of HubSpot's marketing automation or Salesforce's enterprise customization. You're trading best-in-class individual capabilities for good-enough-across-the-board at a lower price point. For many organizations, that trade-off is exactly right.

Making the Decision

The decision framework is simpler than vendors want you to believe. If you need marketing-sales alignment and don't have dedicated RevOps, choose HubSpot. If you need enterprise customization and have the technical resources to support it, choose Salesforce. If you're an agency or need multi-tenant management at aggressive pricing, choose GHL.

Don't choose based on features you might use someday. Choose based on what your team needs today and what they'll realistically adopt. A simple system that everyone uses creates infinitely more value than a sophisticated system that no one uses. Adoption beats capability every time.

The Three-Phase Integration Strategy

Once your CRM is selected, the integration strategy follows a deliberate three-phase progression that mirrors the spreadsheet-first philosophy: prove, then integrate, then scale. Each phase builds on validated results from the previous one, preventing the catastrophic failure mode of deploying everything simultaneously and hoping it works.

Phase 1: Prove the Methodology

Whether you're using spreadsheets or an existing CRM, the first phase focuses entirely on validating that your enrichment data, targeting criteria, and outreach approach generate results. This typically takes three to four weeks.

The discipline of this phase matters more than the tools. Your enrichment data from Chapter 7 flows into a simple daily workflow: review enriched prospects, prioritize based on ICP score and timing signals, execute personalized outreach using the enrichment insights, and track results rigorously. The objective is answering three questions with data, not assumptions: Is our ICP definition generating qualified responses? Is enrichment-powered personalization outperforming generic outreach? Can we achieve consistent response rates above ten percent?

If the answer to all three is yes, you have a validated methodology worth investing in. If any answer is no, you need to iterate on targeting, messaging, or enrichment before adding technology complexity. It's far easier and cheaper to fix a broken process in spreadsheets than in a fully configured CRM with automated workflows that propagate the same broken process at scale.

The signals that you're ready to move to Phase 2 are consistent: response rates above ten percent, a growing pipeline that's becoming difficult to track manually, reps requesting better tools because the methodology is working but the manual process is slowing them down, and management confidence that the ICP and messaging are validated. When your team is asking for the CRM rather than being told to use it, you've earned the right to invest.

Phase 2: Integrate the Infrastructure

With validated methodology in hand, CRM configuration becomes strategic rather than speculative. You know exactly what custom fields you need because you've been tracking that data manually. You know what automation workflows will help because you've been doing those steps by hand. You know what reporting dashboards matter because you've been calculating those metrics in spreadsheets.

The integration architecture connects your enrichment engine, CRM, and outreach tools into a closed loop: Clay enriches prospects, your CRM stores and scores them, your outreach tools execute campaigns, replies flow back into the CRM creating tasks for reps, and the entire cycle generates data for analysis and optimization.

The critical integration points each serve specific purposes. Enrichment-to-CRM moves data either through batch upload or real-time API depending on your volume. CRM-to-outreach uses lifecycle stage or tag triggers to automate campaign enrollment. Outreach-to-CRM captures replies in real time to maintain speed-to-lead advantages. And CRM-to-rep creates task assignments and notifications that ensure no opportunity falls through the cracks.

This phase also includes your email infrastructure decisions. Most CRMs include email functionality, but dedicated outreach platforms solve three problems that CRM-native email typically can't handle at scale: deliverability infrastructure with dedicated IP reputation management, volume capacity beyond CRM-imposed limits, and advanced A/B testing capabilities that enable the optimization methodology from Chapter 11. Tools like Instantly, Smartlead, and Lemlist each have different strengths - Instantly for volume and deliverability, Smartlead for inbox management, Lemlist for advanced personalization. The right choice depends on your volume requirements and personalization strategy.

One non-negotiable rule: never use your primary domain for cold outreach. If cold email gets flagged, it can damage your primary domain reputation and affect all company email delivery. Secondary domains that are recognizable but clearly separate protect your core communications infrastructure while enabling aggressive outbound programs.

The deliverability fundamentals - domain authentication, warmup sequences, sending limits, and reputation monitoring - are technical requirements that your RevOps team or email platform will handle. What matters at the executive level is understanding that deliverability is infrastructure, not an afterthought. Companies that skip proper email setup face months of reputation rehabilitation that effectively shuts down pipeline generation. The investment of two to three weeks in proper warmup and configuration prevents disasters that cost far more in lost pipeline.

Phase 3: Scale What Works

With proven methodology running on integrated infrastructure, scaling becomes a matter of expanding volume while monitoring quality. This is where the compound advantage of systematic AI sales really emerges.

Scaling means expanding your enrichment coverage to larger prospect databases, increasing outreach volume across properly warmed sending infrastructure, adding new personas and segments using the test-first campaign methodology from Chapter 10, and potentially expanding into new markets or product lines using the same systematic approach.

The monitoring discipline during scaling is critical. Response rates should hold within two to three percentage points of your Phase 1 validated results. If they drop significantly, something is degrading - deliverability issues, data quality problems, or message fatigue. Daily monitoring of key health metrics catches these problems before they compound.

The companies that scale successfully treat their AI Sales Engine like a production system, not a marketing campaign. Production systems have monitoring, alerting, maintenance schedules, and performance benchmarks. Marketing campaigns have launch dates and post-mortems. The difference in mindset produces dramatically different long-term results.

Data Hygiene: The Silent Killer of CRM Implementations

Data quality degrades silently and continuously. Without automated hygiene processes, your CRM becomes a graveyard of outdated information that actively undermines your sales team's credibility and effectiveness.

Three automated workflows prevent the most common data quality failures.

The first handles email bounces. When a hard bounce occurs, the system should immediately mark the email as invalid, remove the contact from all active sequences, create a task for re-enrichment through Clay, and if a second bounce occurs on the re-enriched address, flag the contact as "Bad Data - Do Not Contact." This prevents repeated sending to invalid addresses, which is the fastest path to deliverability destruction.

The second manages stale leads. Contacts with no activity and no response after ninety days get reclassified, removed from active sequences, and added to a re-enrichment list that checks for job changes or new email addresses. People change roles, companies evolve, and timing shifts. A prospect who was unresponsive six months ago might be actively evaluating solutions today after a leadership change that your enrichment refresh can detect.

The third automates enrichment refresh. Every ninety days, active prospects get pushed back through your Clay enrichment workflow to update job titles, company data, and timing signals. When significant changes are detected - new role, new company, recent funding, leadership transition - the system alerts the assigned rep. This is how you catch the opportunities that most sales organizations miss: the VP who just moved to a company that fits your ICP perfectly, the prospect company that just raised funding and now has budget for your solution.

These workflows run quietly in the background, but their cumulative impact is enormous. Companies with automated data hygiene maintain ninety-five percent data accuracy. Companies without it watch accuracy decay to sixty or seventy percent within a year, and their campaign performance decays proportionally.

The Eight Integration Pitfalls That Destroy Implementations

Understanding how integration fails helps prevent the predictable mistakes that derail otherwise sound deployments.

Insufficient data mapping is the most common and most preventable failure. Companies rush field mapping during migration, and data ends up in wrong fields, gets truncated, or disappears entirely. Reps can't find enrichment data, personalization hooks appear in the wrong places, and ICP scores are missing or miscalculated. The fix is simple but requires discipline: create a detailed mapping document before any migration, test with a small batch of records first, and have the sales team validate that data appears where they expect it.

The over-automation trap is the most insidious failure because it looks like progress. Companies automate everything - enrollment, sending, follow-up, even responses - and remove the human judgment that makes personalized outreach effective. Response rates decline gradually, prospects comment that outreach "feels automated," and the sales team becomes disconnected from what's actually being sent on their behalf. The fix is starting with semi-automation that requires human triggers at key decision points. Automate the mechanics - data flow, scheduling, tracking - but keep humans in the loop for message quality, personalization review, and strategic decisions about engagement approach.

Poor change management kills more implementations than technical failures. Gartner found that only eleven percent of sales organizations drive commercial success while executing a transformation. The fix is involving the team in the planning process, providing comprehensive training, and starting with simple workflows before introducing advanced features. Chapter 13 covers this in depth, but the principle applies specifically here: phased rollout with quick wins builds adoption momentum that top-down mandates never achieve.

API rate limiting catches teams off guard during high-volume operations. Integration tools hit platform limits, data sync stops randomly, and the team loses trust in the system's reliability. Understanding rate limits for all platforms upfront and building in batch processing for non-critical updates prevents this entirely.

Duplicate records create confusion that compounds over time. Multiple records for the same person with conflicting information make it impossible for reps to trust the data or for management to trust the reporting. Implementing duplicate detection before import and establishing clear master data management rules prevents the problem from starting.

Inadequate security and access control exposes sensitive data or prevents team access to needed information. Planning permission structure before setup, using role-based access control, and regularly auditing permissions prevents both security incidents and productivity losses.

Insufficient testing launches integration without validation, causing failures during live operations. Testing every integration pathway with real data, including error handling and edge cases, prevents the discovery of broken workflows during actual sales activities.

Inadequate monitoring allows integration performance to degrade unnoticed. Setting up dashboards and alerts for integration health catches issues before they impact pipeline generation. Assigning clear ownership for integration maintenance ensures someone is watching.

Executive Takeaways

For the CEO: CRM integration is a strategic decision, not just a technology project. The three-phase approach - prove, integrate, scale - manages risk while maintaining momentum. Total technology stack cost typically runs fifteen hundred to three thousand dollars monthly, significantly less than traditional sales development approaches, with expected ROI of thirty to fifty times in the first year. The key success factor is having an operations owner who can manage the integration - it doesn't need to be a full-time role, but someone needs accountability for system health and data quality.

For the CRO: Phase your rollout to prove value before scaling complexity. The spreadsheet-first discipline prevents the most common failure mode: deploying sophisticated tools with unvalidated methodology. Early metrics targets are response rates above ten percent and meeting booking rates above fifteen percent from responses. Team adoption determines everything - simple systems that teams actually use beat sophisticated systems they ignore. Transition discomfort is real but brief: two to three weeks of adjustment followed by significant productivity gains.

For the CMO: Marketing-sales alignment happens at the CRM integration layer through shared data and unified reporting. Lead scoring must incorporate enrichment data alongside form fills and website behavior to create a complete picture of prospect readiness. Email deliverability is shared responsibility - cold outreach domains must be completely separate from marketing domains to protect sender reputation. The combination of account-based marketing, enrichment data, and CRM automation creates compounding advantages that accelerate over time.

For the CFO: Budget eighteen to thirty-six thousand dollars annually for the complete technology stack versus one hundred thousand-plus for traditional methods. Payback period is typically one to three months based on improved efficiency and pipeline generation. Cost per meeting ranges one hundred to two hundred dollars versus five hundred to one thousand for traditional outbound. And the economics improve with scale: costs grow linearly while output grows exponentially as the systematic approach compounds efficiency gains.

For the CTO/RevOps: This is eighty percent process and twenty percent technology - focus on workflow optimization before tool sophistication. Integration complexity is middleware-level, not custom development. Ongoing maintenance requires three to five hours weekly once running. Technical debt is low with modern SaaS stack versus legacy integrations. Your primary value is ensuring data flows reliably between systems and that the team has the operational support to use the tools effectively.

The companies that successfully integrate these systems aren't just improving their current sales results. They're building scalable revenue engines that become more efficient and effective over time. While competitors are still manually prospecting one contact at a time, your team is operating with complete information advantage, systematic processes, and compound efficiency gains.

Your enrichment engine from Chapter 7 told you what to say. Your integrated CRM and automation systems tell you when to say it, track what happens next, and optimize everything continuously. In Chapter 9, we'll show you how to create bulletproof standard operating procedures that ensure your AI Sales Engine runs consistently regardless of team changes or market conditions.

CHAPTER 9

Crafting Bulletproof SOPs

There's a saying that's been passed down through military training for centuries, originally attributed to the Greek soldier-poet Archilochus: "We don't rise to the level of our expectations. We fall to the level of our training." Navy SEALs put it more bluntly: "Under pressure, you don't rise to the occasion. You sink to the level of your training. That's why we train so hard."

The same principle applies to sales teams - with one critical difference. In the military, training is documented, standardized, and drilled until it's automatic. In most sales organizations, "training" is whatever the last person who was good at the job happened to figure out on their own.

Your AI Sales Engine is generating 340% more pipeline. Your team is hitting every target. Then your best rep gives two weeks' notice. Poached by a competitor. Better offer, more equity, corner office - the usual. You think, no problem. We've got the systems. We've got the enrichment engine, the CRM integration, the campaign sequences. We'll just hire someone new and plug them in.

Except you can't. Because the "system" was never really a system. It was one person who figured out how to make everything work through six months of trial and error - which enrichment signals to trust, which to ignore, how to turn data into a message that sounds human, when to call instead of email, how to handle a bounce without destroying deliverability. None of it was written down. All of it walked out the door with them.

The new reps are drowning. They're copying and pasting AI-generated personalization hooks without reading them - sending emails that reference a "recent merger" to a startup that's never been acquired, or congratulating someone on a funding round that happened eighteen months ago. Your sales manager is spending three hours every morning trying to reverse-engineer what the departed rep did instinctively. Pipeline drops 60% in a month. The CEO starts asking whether the whole AI investment was a mistake.

It wasn't. The investment was sound. What was missing was the layer that turns individual knowledge into organizational capability - standard operating procedures that make the system run the same way regardless of who's sitting in the chair.

James Clear wrote in Atomic Habits: "You do not rise to the level of your goals. You fall to the level of your systems." That's the thesis of this chapter. Technology scales tools. SOPs scale people. And if your processes live in someone's head instead of in documented workflows, you don't have a sales engine - you have a dependency that's one resignation letter away from collapse.

Why SOPs Are Critical for AI Sales Success

Most companies approach AI sales with what I call the "set it and forget it" fantasy. They believe that once you implement the right tools and get them configured properly, the system will run itself and generate predictable results without ongoing management or documentation.

This is like expecting a Formula 1 race car to drive itself just because it has sophisticated technology. The technology amplifies capability, but it doesn't replace the need for systematic processes to operate it effectively.

I've seen this play out dozens of times. A company spends four months building a beautiful AI sales engine. They have Clay enriching thousands of prospects weekly. They have HubSpot automating sequences and scoring leads. They have Instantly managing email deliverability across multiple domains. The technology is flawless. Six months later, their results are worse than before implementation.

What happened? They automated everything but systematized nothing. Different reps were using the enrichment data differently. Some were personalizing heavily, others were ignoring it completely. The manager couldn't tell which approaches were working because there was no standard process to measure against. When reps succeeded, no one knew why. When they failed, no one knew how to fix it.

The company had confused automation with systematization. Automation is technology doing things without human intervention. Systematization is humans doing things consistently according to documented processes. You need both, but systematization comes first.

What Happens Without SOPs - and What Happens With Them

Let me show you what organizational chaos actually looks like when you have powerful technology but no documented processes.

The first thing that breaks is knowledge transfer. Every successful rep develops their own personal system for using AI tools effectively. They figure out shortcuts, identify patterns, and develop instincts about what works. But this knowledge stays locked in their individual experience instead of becoming organizational capability. When they succeed, the company can't replicate their approach with other reps. When they leave - and they always eventually leave - their knowledge walks out the door with them.

The second casualty is execution consistency. Without documented processes, every rep uses enrichment data differently. Some spend fifteen minutes per prospect crafting thoughtful emails. Others spend two minutes sending generic messages with a few personalization tokens. Some prioritize timing signals heavily. Others ignore them completely. The result is wildly inconsistent performance that management can't analyze or improve. You end up with one rep hitting one hundred and fifty percent of quota while another hits sixty percent, using the same technology and targeting the same types of prospects.

Then there's the onboarding problem. New hires have to reinvent every process from scratch. They spend weeks figuring out how to interpret ICP scores, how to use personalization hooks effectively, and how to prioritize their daily activities. What should take days takes months, and even then they might never reach the same level of effectiveness as the original team.

And without a documented baseline, optimization becomes impossible. You can't improve what you can't measure, and you can't measure what isn't standardized. Teams end up making random changes hoping something will work better, instead of making data-driven improvements to documented processes.

The final symptom is decay. Even successful implementations gradually deteriorate without documented processes. Reps develop bad habits, take shortcuts that seem efficient but reduce effectiveness, and drift away from the approaches that originally generated results. Month one after implementation: fantastic results. Month six: mediocre results. Month twelve: worse than before implementation. The technology hasn't changed. The processes have decayed through inconsistent execution and tribal knowledge loss.

Companies with properly documented processes see completely different outcomes, even using identical technology. When processes are documented, every rep follows the same workflow for reviewing enrichment data, crafting outreach, and managing follow-up sequences. This creates predictable baselines that management can measure, analyze, and improve systematically.

New reps can follow proven workflows immediately, reducing time-to-productivity from months to days. Clear expectations at each step enable managers to identify exactly where performance breaks down and provide specific coaching instead of generic feedback. And when you do test improvements, you can change one element of the workflow, measure the results, and either implement the improvement across the team or revert to the previous process.

The Three Types of SOPs Your AI Sales Engine Needs

Every AI Sales Engine needs three categories of standard operating procedures to function effectively. Understanding this framework prevents the common mistake of documenting some processes while leaving critical gaps.

Daily Operations SOPs define what sales reps do every day to generate consistent results. These include prospect research workflows, outreach creation processes, follow-up sequences, and CRM management tasks. These are the most critical SOPs because they directly impact revenue generation.

Management SOPs define what sales leaders do to monitor performance, coach teams, and optimize results. These include pipeline review processes, performance analysis workflows, team coaching procedures, and system maintenance tasks. These SOPs ensure that leadership can scale their effectiveness across growing teams.

System Maintenance SOPs define how to keep the technology infrastructure running smoothly. These include data enrichment refresh procedures, email deliverability monitoring, CRM hygiene workflows, and integration troubleshooting processes. These SOPs prevent technology problems from derailing sales results.

All three types work together to create systematic success that doesn't depend on individual heroics or tribal knowledge.

The Daily Top 25 Review: Your Most Important SOP

The most important SOP in your entire system is the one that determines how sales reps start each day. Get this wrong and everything else fails.

I've seen this dynamic play out clearly: one rep consistently hits one hundred and forty percent of quota while others on the same team struggle to reach eighty percent. The difference isn't talent, territory, or luck. The high performer has developed a systematic approach to prioritizing prospects each morning, while everyone else is cherry-picking accounts that seem easier without any strategic framework.

The high performer arrives early and spends fifteen minutes reviewing the enriched Top 25 list. He reads every personalization hook, checks timing signals, and ranks prospects based on a framework he's developed over months. He knows exactly who to contact first and why. The other reps glance at the list, pick a few names that look familiar or easy, and start sending emails without any systematic approach. They're working just as hard but getting completely different results.

When you document the high performer's process and have the entire team follow it, average quota attainment jumps dramatically within weeks. Same prospects, same enrichment data, same technology. Different process.

Here's what that process looks like when properly documented. The manager's prep work begins each morning with a refresh of ICP scoring in Clay, which runs automatically but needs manual verification. While the system updates scoring based on new enrichment data overnight, the manager reviews results to catch obvious errors or data anomalies that might skew prioritization. The refreshed Top 25 list gets exported to a shared format and posted in the team's communication channel at a consistent time with key highlights - which companies just raised funding, which have new leadership, which contacts changed roles.

The rep's review starts immediately. Open the Top 25 list and spend exactly fifteen minutes reviewing every prospect systematically, not randomly. For each prospect, read the AI-generated personalization hook completely. Don't just scan it. Ask yourself whether it makes sense, whether it's accurate, whether it's recent, and whether it would grab your attention if you received it. Review the enrichment data for context - company size for decision-making complexity, recent signals for urgency, technology stack for conversation starters, timing indicators that suggest evaluation mode.

Then make channel decisions. Email works best when you have a strong personalization hook and verified email address. Phone calls work when the personalization hook is weak but the company fit is strong. LinkedIn outreach works when you have mutual connections or shared experiences.

The critical prioritization step: choose eight to ten focus accounts for the day's outreach. Don't pick more than ten - you won't do them justice. Don't pick fewer than eight - you're not maximizing your opportunity. Priority one goes to prospects with high ICP scores combined with recent timing signals. Priority two covers high ICP scores with verified contact information but no specific timing triggers. Priority three includes medium ICP scores with unusually strong personalization hooks. Skip prospects with low scores, unverified contact information, or weak hooks that you can't strengthen.

A manager spot check at the start of the day with randomly selected reps keeps the process honest. Ask specific questions: who are you targeting today and why, what's compelling about your top prospects, what timing signals are you leveraging. Listen for understanding versus mere compliance. You want reps who can explain their prioritization, not just confirm they looked at the list.

Converting Enrichment Data Into Human Outreach

The gap between having enrichment data and using it effectively destroys more AI sales implementations than any technical failure.

Here's the pattern I see constantly: reps send emails like "Hi John, I noticed your company is in the construction industry and thought you might be interested in our software that helps construction companies be more efficient." This email is technically personalized - it uses the prospect's name and industry. But it's completely generic and could have been sent to any executive in the database. Meanwhile, the rep has access to enrichment data showing that John's company just raised twelve million dollars, hired fifteen new employees in the past sixty days, and recently implemented Salesforce after years of using spreadsheets. None of this compelling context made it into the outreach.

When you ask reps why they aren't using the enrichment data, the answer is almost always the same: "It takes too long to read through all that information. I can send forty emails using templates in the time it takes to write eight personalized ones." They're optimizing for quantity over quality and getting predictable results - two percent response rate on forty emails per day versus a top performer who gets sixteen percent response rate on twelve highly personalized emails per day.

The solution is a systematic process that converts enrichment data into compelling outreach in about eight minutes per prospect. The process has a specific rhythm that becomes automatic with practice.

The first two minutes go to pulling and reading enrichment data. Open the prospect's record and read every custom field populated by your enrichment process. Don't glance at the data - read it for understanding. Company revenue and employee count tell you about decision-making complexity. Industry and sub-industry point to specific pain points. Recent growth indicators suggest budget availability. Technology stack opens specific conversation angles.

One minute goes to human validation. Make a quick LinkedIn visit to verify the contact's current role and recent activity. Take a thirty-second glance at the company website to confirm they're still in business and operating in the same market. This step prevents disasters. I've seen reps congratulate prospects on funding rounds that happened two years ago and job changes that never occurred. These mistakes damage credibility permanently. If any enrichment data seems wrong or outdated, update the CRM record immediately. Don't send outreach based on bad information.

One minute goes to reviewing and refining the AI-generated personalization hook. Read it completely and apply a two-question test: does this sound like something a human would write, and is this information accurate and recent? If both answers are yes, use the hook or modify it slightly to match your voice. If either answer is no, write your own hook using the enrichment data you've reviewed.

The difference between AI-generated and human-refined personalization matters enormously. An AI-generated hook might say "I saw your company raised fifteen million dollars last month." A human-refined version says "Congrats on the Series B - saw the announcement in TechCrunch last week. That kind of growth capital usually means scaling the team quickly." The human version demonstrates actual awareness rather than automated data insertion.

Three to five minutes go to drafting the actual outreach. For emails, use a structure that flows naturally: a personalized opening line using an enrichment insight, a brief problem identification based on their situation, a one-sentence solution presentation with specificity, and a simple call-to-action with a low commitment threshold. For call preparation, develop talking points rather than scripts - an opening reference to the personalization hook, a problem statement based on their stack or position, qualification questions about their current situation, and prepared responses to likely objections.

The final minute is quality control. Read your email out loud before sending. Ask yourself honestly whether you would respond to this message. Check for basic errors - correct company name, correct person name, accurate recent information, clear call-to-action. If it feels templated or generic, revise until it feels genuine. Then send immediately while context is fresh and log the activity in your CRM with specific details, not generic notes like "sent email." Include context that will help with follow-up: what you referenced, what angle you took, when to follow up.

Eight to ten minutes per prospect compared to forty-five minutes of manual research or two minutes of templated outreach. That's the sweet spot where quality and volume intersect.

Keeping Your Intelligence Fresh: The Weekly Enrichment Refresh

Stale enrichment data doesn't just reduce effectiveness - it can actively damage your reputation and relationships.

The cautionary tale goes like this: a sales rep reaches out to congratulate an executive on his new role as Chief Revenue Officer, referencing his recent promotion and asking about growth initiatives for the new year. The problem is the executive left that company four months earlier to join a competitor. The new CRO at the original company forwards the email to his team with the comment, "This is why I don't trust vendors who claim to do research." That single mistake based on stale data costs the company a major opportunity and damages their reputation with an entire prospect organization. The executive who received the mistaken outreach mentions it to peers at industry events, creating negative word-of-mouth that affects credibility for months.

This disaster is completely preventable with systematic data refresh processes.

The weekly refresh combines automated and manual review in a specific cadence. Your Clay automation triggers overnight each weekend and systematically re-enriches all contacts marked as active prospects in your CRM - typically five hundred to two thousand records depending on your pipeline volume. The system checks for job title changes, company updates including funding and acquisitions and leadership changes, technology stack changes, and new intent signals. When changes are detected, it updates CRM fields automatically and flags significant changes for manual review.

Monday morning, the manual review begins. Work through each flagged item systematically. For contact job changes, determine whether the person moved to a new company or changed roles within the same organization. If they moved companies, update their CRM record and evaluate whether the new company fits your ICP. If it's a good fit, create a new prospect record and assign it to the appropriate rep with context about the previous relationship - a warm outreach opportunity.

For company funding events, update ICP scores immediately. Alert the sales team through your communication channel with context about why the funding matters and how it changes the prospect's position on the priority list. For new decision-makers, add the contact and research their background for personalization opportunities. For technology stack changes, reassess relevance to your solution - if they adopted a complementary technology, outreach should focus on integration opportunities; if they removed a competitive tool, prioritize immediate outreach about the evaluation process.

After the manual review, re-score the Top 25 using refreshed enrichment data and compare this week's list to last week's. Companies moving up in ranking often indicate increased receptiveness due to timing changes. Companies moving down might indicate reduced fit or changing priorities. Send the updated list to the team with the three to five most significant changes highlighted and context about why they matter.

The key metrics for your refresh process are data freshness - what percentage of active contacts were updated this week, targeting ninety-five percent or higher - and flagged change completion rate, which should be one hundred percent. Running automated refresh without reviewing flagged changes wastes the intelligence you're paying for.

Protecting Your Deliverability: Bounce Handling as Infrastructure

Email deliverability affects every aspect of your AI Sales Engine performance, and bounced emails are leading indicators of deliverability problems that can cascade quickly if ignored.

The pattern is always the same. A company ignores bounced emails for a few months, assuming bounces are just part of cold outreach. They continue sending emails to invalid addresses. Their deliverability rate gradually declines as ISPs flag their domains for poor list hygiene. By the time they notice, all their emails are landing in spam folders - including emails to valid addresses. Response rates crater not because messaging got worse, but because prospects aren't seeing the emails at all. Rehabilitating sender reputation takes weeks of careful work, during which pipeline generation essentially stops.

This is completely preventable with systematic bounce handling.

The process starts with real-time bounce notification through webhook integration between your email tool and CRM. When your email platform detects a bounced email, it immediately triggers a CRM update and creates a task with a four-hour response deadline. The deadline exists because bounces indicate data quality problems that affect deliverability reputation. Fast response prevents individual bounces from becoming systematic degradation.

Within four hours of the bounce notification, triage the bounce type. Hard bounces mean emails that don't exist or domains that reject mail permanently - these require immediate action because continued sending to invalid addresses damages sender reputation rapidly. Soft bounces indicate temporary delivery problems like full mailboxes or server issues - these addresses might become deliverable later but require careful handling. Spam complaints are the most serious type, indicating prospects actively marked your email as spam, and require immediate permanent removal from all lists.

For hard bounces, initiate re-enrichment immediately. Run the bounced contact through your enrichment tools using different data providers in a waterfall approach. If your primary provider failed, try a second, then a third. Different providers have different data sources and update schedules. If re-enrichment finds a new email address, verify it through a verification service before resuming outreach. If no new email is found, either find a different decision-maker at the same company or mark it for revisit in ninety days and remove from active outreach.

Update the CRM record with detailed notes on what happened, what was tried, and what the current status is. Remove the contact from all active email sequences to prevent additional bounces. Don't give up too quickly on high-value prospects - I've seen companies successfully reach prospects on the fourth different email address attempt when the first three bounced.

The target metrics are time to fix bounces under four hours, re-enrichment success rate of sixty to seventy percent, and monthly bounce rate under two percent.

Training Your Team to Actually Use What You've Built

The most sophisticated technology in the world is worthless if your team doesn't know how to use it effectively. I've seen companies spend tens of thousands of dollars implementing a perfect AI sales engine and then spend zero dollars training their team how to use it.

Six months later, the team is using about ten percent of the system's capabilities and getting worse results than before implementation. The enrichment data is there, but reps are ignoring it. The personalization hooks are generated, but reps are sending generic templates. The timing signals are identified, but no one understands what they mean or how to use them.

When you ask the CEO why they didn't invest in training, the answer is always some version of "these tools are supposed to be intuitive." This is like buying Formula 1 race cars and expecting your team to drive them effectively without instruction because cars should be intuitive to drive.

Enrichment data training needs to be systematic and practical. Start with a side-by-side demonstration: show a blank contact record with just name and email, then show the same contact after enrichment with company size, technology stack, recent funding, personalization hooks, and timing signals. Explain that this represents research your sales team used to spend two to three hours doing manually for each prospect.

Walk through each custom field with specific examples. The ICP Fit Score shows how well this prospect matches your ideal customer profile based on data analysis of your best customers -

scores above one hundred indicate strong fit, eighty to ninety-nine indicate moderate fit, below eighty suggests poor fit or missing data. Technology stack information tells you what tools they're using, which affects both fit assessment and conversation starters. Recent signals indicate timing opportunities. Personalization hooks are AI-generated conversation starters based on recent company activity - starting points, not finished products.

The practice component is critical. Give each trainee three enriched contact records with different characteristics and task them with writing personalized email openers using the enrichment data. Have them share their work with the group. Look for evidence that they're using specific enrichment insights rather than generic industry information. Good personalization references specific recent events, technology context, or company situations. Poor personalization uses broad generalizations that could apply to any similar company.

Cover common mistakes explicitly. Copying AI-generated hooks verbatim without verification leads to outreach based on outdated or incorrect information. Ignoring enrichment data and sending generic emails wastes your entire data investment. Not verifying data accuracy leads to embarrassing emails based on wrong information. Using outdated signals - congratulating someone on funding that happened eight months ago - makcs you look automated rather than attentive.

Certification should require completing the training, writing real personalized emails based on actual prospect data, having those emails approved by a manager before sending, and demonstrating understanding through verbal explanation. This isn't paperwork - it ensures quality control and prevents expensive mistakes from undertrained team members.

Reinforce with weekly coaching sessions where managers review sample outreach and provide feedback on enrichment usage, monthly team meetings to discuss new features and share success stories, and quarterly refresher training as tools and data sources evolve.

SOP Maintenance: Living Documents, Not Shelf Decorations

Standard operating procedures aren't "set it and forget it" documents. They're living systems that must evolve as your tools change, your processes improve, and your team scales.

The failure pattern looks like this: a company documents comprehensive SOPs when they implement their AI Sales Engine. Two years later, half the SOPs reference tools they no longer use and processes that have evolved significantly through team experience. New reps follow outdated procedures that confuse rather than help them. The Top 25 review process has been optimized from thirty minutes to fifteen minutes through team experience, but the SOP still describes the old approach. The enrichment refresh procedure references features that have been updated with better functionality. The email sequencing workflow describes capabilities that have been replaced with better automation. The result is systematic confusion rather than systematic success.

The fix is a quarterly review process. Assign SOP ownership to RevOps in partnership with sales leadership - this isn't a task you can delegate to junior team members because it requires strategic understanding of how processes connect to results. Schedule quarterly review sessions with a specific agenda: what's working as documented, what's changed since the last update, what new processes need documentation, what SOPs are no longer relevant, and how SOPs can be improved based on team feedback and results data.

Gather feedback systematically from the team members who use SOPs daily. Which SOPs are unclear or confusing? What processes do they do differently than what's documented? What new workflows have they developed that should be documented? Where do new hires struggle most during onboarding?

Update SOPs based on collective wisdom rather than individual preferences. If multiple team members independently developed the same process improvement, document it as the new standard. If only one person uses a different approach, understand why before deciding whether to update the SOP or coach the individual.

Implement version control. Date all revisions, maintain previous versions for reference, and communicate changes clearly to the entire team. Don't just update documents quietly - announce changes and explain why they matter. Track compliance through weekly spot checks rather than predictable audits. Look for correlation between SOP compliance and performance. High-performing reps who ignore SOPs might have discovered improvements worth documenting. Low-performing reps who ignore SOPs need coaching on process adherence.

The goal is continuous improvement through documented, tested processes. The companies that do this well don't just avoid the hero dependency trap - they create sustainable competitive advantages that compound over time. While competitors rely on individual heroics that can't be scaled or replicated, they operate with systematic processes that get more efficient and effective as teams grow.

Chapter 9 Executive Takeaways

For the CEO: SOPs aren't bureaucracy - they're how systems beat heroics every time. Your AI Sales Engine investment only delivers ROI if processes are documented and followed consistently. The expected impact is dramatic: sixty percent reduction in new rep ramp time, elimination of key-person dependency, and systematic scalability. The investment required is forty to sixty hours for initial documentation and five to ten hours quarterly for maintenance. ROI typically runs ten times the investment in the first year through faster scaling and reduced key-person risk.

For the CRO: Daily operations SOPs enable consistent execution regardless of who's selling, protecting you from the hero dependency trap. Enrichment-specific SOPs protect your data investment by ensuring information is used effectively rather than ignored. Enforcement matters critically - SOPs only work if consistently followed and measured. Track SOP compliance rates against performance results to identify coaching opportunities and process improvements.

For the CMO: Marketing-sales SOPs create handoff clarity and eliminate confusion about lead qualification and routing. Enrichment refresh SOPs keep your marketing database current for campaign effectiveness. Account-based marketing requires systematic coordination SOPs between marketing and sales teams. Document successful processes and scale through systems rather than depending on individual relationships.

For the CFO: SOPs reduce key-person risk by preventing critical knowledge from being trapped in individual heads. Training costs drop seventy percent when processes are documented rather than learned through trial and error. Scalability improves dramatically - you hire and onboard faster with lower failure rates. SOPs create audit trails that enable accountability and measurable performance improvement.

For the CTO/RevOps: You own SOP creation and maintenance as core infrastructure, not optional documentation. Eighty percent of sales issues are process problems, not technical problems. SOPs make troubleshooting systematic by clearly defining what should happen versus what is happening. Living documentation prevents technical debt from process drift and keeps teams aligned with system capabilities.

The companies that document their AI Sales Engine processes systematically don't just avoid the hero dependency trap - they create sustainable competitive advantages that compound over time. While competitors rely on individual heroics that can't be scaled or replicated, they operate with systematic processes that get more efficient and effective as teams grow.

Your enrichment engine from Chapter 7 and your integrated CRM automation from Chapter 8 are powerful tools. But tools don't create consistent results. People following systematic processes create consistent results. SOPs are how you transform AI technology into predictable revenue generation that works regardless of team changes, market conditions, or competitive pressures.

In Chapter 10, we'll show you how to launch high-impact AI campaigns that leverage all the systematic foundations you've built - enrichment, integration, and processes - to generate pipeline velocity that competitors can't match.

PHASE 3: ACTIVATE

Chapter 10:

Launching High-Impact AI Campaigns

100 prospects. Three approaches. Two weeks.

That's all it takes to know whether your AI Sales Engine actually works.

Not a six-month rollout. Not a quarter-long "pilot program" with a 40-page post-mortem. Not a company-wide campaign launch that bets your entire pipeline on messaging nobody's tested. One hundred carefully selected prospects from your ICP, three different approaches running simultaneously, and fourteen days of data.

At the end of two weeks, one of three things will be true. One approach bombed - low single-digit response rates, no meetings, dead air. One approach was decent - respectable numbers, a few meetings, signs of life. And one approach is clearly winning - double-digit response rates, meetings booked, prospects engaging, pipeline forming. You kill the losers, scale the winner to your full database, and you're generating real revenue within a month of launch.

That's it. That's the entire activation methodology.

CEOs hate hearing this because they've been conditioned to believe that campaign launches require months of preparation, cross-functional alignment meetings, creative reviews, brand guidelines committees, and seventeen rounds of stakeholder feedback on subject lines. They want the launch to be perfect. They want guarantees before they press send.

Here's the truth: there are no guarantees. There's only data. And you can't get data without sending something to someone. The companies that win aren't the ones with the perfect campaign - they're the ones that tested three imperfect campaigns, found out what worked in two weeks, and scaled it while their competitors were still arguing about copy.

Everything you've built in the last nine chapters - the ICP Blueprint, the intelligence layers, the enrichment engine, the CRM integration, the SOPs - has been leading to this moment. The infrastructure is ready. The processes are documented. The data is flowing. Now it's time to find out what actually resonates with your market, and the fastest way to do that is to stop planning and start testing.

From Infrastructure to Activation

Most companies approach campaign launches backward. They start with creative brainstorming sessions about messaging, positioning, and offers. They debate subject lines and call-to-action language. They spend weeks crafting perfect copy and building elaborate sequences. Then they launch to their entire database and hope for the best.

When results are mediocre, they don't know why. Was it the targeting? The message? The timing? The offer? They can't isolate variables because they changed everything at once and tested nothing systematically.

The pattern is always the same. A company spends two months developing what they call their "perfect campaign." They craft messages for different personas, build five-touch email sequences, create custom landing pages, and coordinate multi-channel outreach across email, LinkedIn, and phone. They launch to thousands of prospects simultaneously and generate a response rate that barely covers the cost of the campaign infrastructure.

When you analyze what went wrong, the messaging assumed prospects understood their core problem. But the enrichment data tells a different story. Most of the database consists of prospects who don't know they have a problem yet. The campaign was targeting the already-convinced while ignoring the unaware.

The campaign failed not because it was poorly executed, but because it was based on assumptions rather than testing. This is the backward campaign planning problem that destroys most AI sales initiatives. Companies optimize for creative perfection instead of market validation. They assume they understand their audience instead of testing their assumptions. They launch big instead of learning fast.

The alternative is what I call the test-first campaign framework, and it transforms campaign development from expensive guesswork into systematic optimization.

The Test-First Campaign Framework

The fundamental principle is simple: never launch to your entire database until you've proven your approach works with a smaller, controlled group.

This seems obvious, but most companies resist it because testing feels slower than launching. They're worried about missing opportunities while they test, so they prefer to take bigger risks with bigger potential rewards.

This thinking is backwards. Testing isn't slower than launching - it's faster than failing. The data supports this conclusively. A Backlinko analysis of twelve million outreach emails found that personalized messages receive 32.7% more replies than generic ones, but the key finding was that the type of personalization matters enormously. Companies that test which personalization approach resonates with their specific market before scaling consistently outperform those that guess.

The test-first framework applies this principle systematically. Every campaign starts with a two-week test of one hundred prospects across three different approaches, with systematic measurement of response rates, meeting quality, and pipeline generation. Over eighteen months of disciplined testing, you'll evaluate dozens of different campaign approaches. Most will fail to meet your response rate threshold and get killed before scale. The handful that succeed get scaled with complete confidence because you've already proven they work.

The result: cost per meeting drops sixty to seventy percent compared to spray-and-pray approaches, while meeting quality and pipeline conversion rates improve dramatically. You fail faster and cheaper. But when you find something that works, you scale it with confidence.

The test-first framework isn't about being cautious - it's about being systematic. You take the same risks, but you take them intelligently with controlled downside and measured upside.

What Makes AI Campaigns Different from Traditional Campaigns

Traditional outbound campaigns are essentially broadcast marketing. You craft one message, segment your database into broad categories, and send roughly the same content to everyone hoping that statistical averages will generate enough responses to justify the cost.

The messaging is generic because it has to work across thousands of different companies, industries, and situations. You can't personalize effectively at scale using manual processes, so you default to the lowest common denominator: broad pain points, generic value propositions, and one-size-fits-all solutions.

AI-powered campaigns are fundamentally different because they enable hyper-personalization at scale. Instead of broadcasting the same message to everyone, you can create unique, relevant, contextual outreach for each prospect based on specific information about their company, situation, and timing.

This isn't just about inserting their name and company into a template. It's about crafting messages that demonstrate genuine understanding of their specific situation and present solutions that address their particular challenges.

The difference is dramatic. Traditional campaign messaging sounds like this: "Hi First Name, I help companies like Company Name improve sales efficiency through better processes and technology. Many companies in your industry struggle with manual processes that slow down growth. Our solution has helped similar companies increase productivity by 40%. Worth a brief call?"

This message could be sent to any prospect in any industry with minimal customization. It's not wrong, but it's completely generic and forgettable.

AI-powered campaign messaging sounds like this: "Hi Sarah, I noticed TechCorp just posted three sales roles and you're using Salesforce but don't have any automation tools connected to it yet. Most VPs of Sales I talk to at Series B companies scaling from 8 to 15 reps run into the same challenge: keeping everyone consistent without drowning in coaching time. We've helped similar companies automate their sales workflows while maintaining the personal touch that closes deals. Worth a 15-minute call to see if there's a fit?"

This message demonstrates specific knowledge of Sarah's company situation, hiring plans, technology stack, and likely challenges. It's not just personalized - it's relevant and timely.

The real-world results from companies using this approach are striking. Rippling, the workforce management platform, doubled their cold email performance year-over-year after implementing Clay-powered enrichment and personalization at scale. Their growth team described Clay as their "secret weapon" for building an experimentation-driven GTM motion that iterates on ideas and scales what works. Sendoso, the direct mail and gifting platform, generated over one million dollars in pipeline and 10x'd their outbound productivity - one SDR now functions like a full team - by automating personalized research and outreach through enrichment workflows.

The enrichment data from Chapter 7 makes this level of personalization possible at scale. Instead of guessing about prospect situations, you know their technology stack, recent funding events, hiring patterns, and company news. Instead of assuming what they care about, you can reference specific signals that indicate timing and receptiveness.

The CRM integration from Chapter 8 makes this level of personalization operationally feasible. Instead of manual research and custom email creation, you can use automated workflows that pull enrichment data into email templates and sequence logic that adapts based on prospect characteristics.

The SOPs from Chapter 9 make this level of personalization consistently effective. Instead of hoping reps will use the enrichment data properly, you have documented processes that ensure every outreach message leverages specific insights appropriately.

The combination transforms campaigns from generic broadcasts into targeted conversations that feel like genuine business development rather than mass marketing.

The 2-Week Campaign Test Framework

Two weeks is the perfect campaign test window because it balances statistical significance with speed and cost control.

Two weeks is long enough to gather meaningful response data from your test cohort. Most prospects will see your first email within twenty-four to forty-eight hours, and those who respond typically do so within three to five days of receiving relevant outreach. A fourteen-day window captures both immediate responses and delayed responses while giving your sequences time to complete their full cycle.

Two weeks is short enough to limit risk and cost if an approach doesn't work. You're testing with a small subset of your database, so failed approaches don't burn your best prospects or damage your sender reputation. The financial investment is minimal compared to full-scale campaign launches.

Two weeks is fast enough to maintain momentum and team energy. Longer test periods lose urgency and allow teams to get distracted by other priorities. Shorter periods don't provide enough data to make confident scaling decisions. The discipline matters - run two-week tests every month for a year, building each test on learnings from previous tests, and you create systematic improvement in campaign effectiveness that compounds over time.

Selecting Your Test Cohort

The prospects you choose for testing determine the validity of your results, so selection methodology matters more than most companies realize.

Don't test on your best prospects because you want to save them for approaches you've already proven effective. Your highest-scoring ICP prospects, companies with recent timing signals, and warm leads from previous interactions should be reserved for scaled campaigns using validated messaging.

Don't test on poor-fit prospects because results won't indicate real potential. Testing messaging on companies that don't match your ICP or prospects who are clearly not decision-makers will give you artificially low response rates that don't predict scalability to better-fit prospects.

Use middle-tier prospects from your Top 25 rotation - companies that score eighty to ninety-five on your ICP fit scale, have verified decision-maker contact information, and represent the bulk of your addressable market. These prospects provide the most accurate indication of how your campaigns will perform when scaled to your full database.

Target exactly one hundred prospects for your test cohort. This number is large enough to achieve statistical significance - you need at least fifteen responses to draw meaningful conclusions about approach effectiveness. It's small enough to be manageable and affordable. It's round enough to make percentage calculations easy during analysis.

Divide your one hundred prospects evenly across the three approaches you're testing: thirty-three prospects for Approach A, thirty-three for Approach B, and thirty-four for Approach C. Assign prospects to approaches randomly rather than trying to match approaches to specific prospect characteristics - you want to test message effectiveness, not targeting precision.

The Three-Approach Testing Strategy

Every campaign test should evaluate three fundamentally different messaging approaches that address different stages of problem awareness and buying readiness.

Approach 1: Problem-Agitation

This is traditional pain-point focused messaging that assumes prospects understand they have a problem and are actively looking for solutions. The approach leads with pain point identification, agitates the consequences of not solving the problem, and positions your solution as the resolution.

Structure: "You're probably dealing with a specific problem. This creates negative consequences that impact business outcomes. We solve this by our approach so you can achieve positive outcomes."

Example: "Hi John, most VPs of Sales at companies using Salesforce without automation tools tell me they're spending two to three hours daily on data entry instead of selling. This usually means your team is hitting about sixty percent of quota potential while burning out from administrative tasks. We integrate AI automation directly with Salesforce so your reps can focus on what they do best - closing deals. Worth a fifteen-minute call to show you how we've helped similar companies increase quota attainment by forty percent?"

This approach works when prospects are already aware they have a problem and are actively evaluating solutions. It's effective for companies showing intent signals like downloading comparison guides, visiting competitor websites, or posting job openings for roles related to your solution area. Use enrichment data to reference specific technology stack gaps, recent hiring patterns that indicate scaling challenges, or company news that suggests operational pressure.

Approach 2: Insight-Education

This is thought leadership focused messaging that assumes prospects don't realize they have a problem or don't understand the full implications of their current situation. The approach leads with industry insights, educates about trends or challenges, and positions your solution as forward-thinking preparation.

Structure: "Most leaders at your level don't realize that an industry trend is creating an emerging challenge. Companies that adapt early typically see competitive advantage. Here's how a specific tactic positions you ahead of competitors."

Example: "Hi Sarah, most CROs don't realize that AI adoption in sales has nearly doubled in the last eighteen months - Salesforce research shows forty-three percent of sales reps now actively use AI, up from twenty-four percent the year before - creating a significant competitive gap. Companies still using manual prospecting are falling behind in pipeline generation compared to AI-powered competitors. The companies adapting fastest are using AI to automate research and personalization while keeping the human touch in conversations. Worth a brief call to share what we're seeing in the market and how leading companies are staying ahead?"

This approach works when prospects aren't actively seeking solutions but need education about emerging trends. The Salesforce data point is powerful because it's verifiable and represents a real market shift that your prospects can confirm independently. Use enrichment data to reference their current technology stack position relative to industry trends, their company size and growth stage relative to optimization opportunities, or their market position relative to competitive threats.

Approach 3: Timing-Signal

This is urgency focused messaging that assumes prospects are experiencing recent changes that create evaluation windows for new solutions. The approach leads with acknowledgment of their situation, connects the change to likely implications, and positions your solution as timely assistance.

Structure: "I noticed a recent change at your company. This usually means likely implications and creates opportunities for strategic improvement. Companies in similar situations typically see specific benefits by addressing this challenge quickly."

Example: "Hi Michael, I saw that DataCorp just raised fifteen million dollars in Series B and you've posted five sales roles in the past month. Most VPs of Sales I talk to in similar growth situations run into the same challenge: maintaining consistency and quality while scaling the team quickly. The companies that handle this transition best usually implement systematic sales processes before the team grows, rather than trying to fix things after they're already struggling. Worth a call to share what's worked for other Series B companies scaling their sales teams?"

This approach works when enrichment data reveals recent changes like funding events, leadership transitions, technology implementations, or market entries that typically trigger evaluation of operational improvements. Use enrichment data to reference specific timing signals while connecting them to likely implications and strategic opportunities.

Building the Test Sequences

Each approach gets a four-touch sequence over fourteen days, designed to test message resonance while providing enough follow-up to capture delayed responses. Research from Apollo.io, which analyzed data from hundreds of millions of emails sent, found that follow-up emails can lead to booking four times the meetings compared to single-touch campaigns. Don't send one email and declare the approach dead.

Email 1 (Day 0): Initial Personalized Outreach. This is your primary test message using the approach-specific structure and messaging style. Use enrichment data from Clay to personalize the opening, reference specific company or contact details, and demonstrate genuine research.

Email 2 (Day 3): Value-Add Content. The second email provides useful information regardless of whether they're interested in your solution, building credibility while maintaining the conversation thread. Share a relevant resource, insight, or data point that demonstrates expertise without pushing for a meeting.

Email 3 (Day 7): Different Angle. The third email approaches from a different perspective while maintaining the same core approach methodology. If you led with a problem, now share how companies in similar situations have addressed it. If you led with an insight, share new data that reinforces the trend. If you led with a timing signal, connect the change to a different implication they may not have considered.

Email 4 (Day 14): Breakup. The final email acknowledges the lack of response while leaving the door open for future engagement. "I've reached out a few times about this topic. I know you get hundreds of emails, so I'm assuming either the timing isn't right or this isn't a priority right now. Should I close your file, or would you prefer I check back in a few months?" This email consistently generates the highest response rates in most sequences because it removes pressure and gives prospects permission to engage on their terms.

Measuring Test Results

Test measurement must focus on business outcomes rather than vanity metrics. Track response rate as percentage of prospects who reply to any email in the sequence, not just opens or clicks. Measure meeting booking rate as percentage of responses that result in scheduled meetings. Evaluate meeting quality by tracking show rate, ICP fit confirmation, and progression to next stage.

Calculate cost per meeting including enrichment costs, tool costs, and time investment. Determine statistical significance by ensuring at least fifteen total responses across all approaches before drawing conclusions.

Here's what decisive test results typically look like: one approach generates single-digit response rates with almost no meetings booked. Another generates moderate response rates with a couple of meetings. And the third approach is clearly winning - double-digit response rates, meetings booked, prospects engaging, pipeline forming. You kill the losers, scale the winner, and you're generating real revenue within a month of launch.

Scaling Your Winning Campaign

The decision to scale requires clear success criteria and systematic rollout methodology. Don't scale until you have a clear winner with response rates above fifteen percent and consistent meeting quality. Don't scale too slowly because market windows close and competitive advantages erode. Don't scale too quickly because deliverability issues and quality degradation can destroy successful campaigns.

Phase 1: Controlled Scale

Expand to five hundred prospects using the winning approach while monitoring key metrics closely. Response rates should hold within two to three percentage points of test results. Meeting quality should remain consistent with your test cohort. Deliverability metrics should stay strong with bounce rates under two percent and spam complaints under 0.1 percent.

The temptation is always to skip this phase and go straight to full scale. Resist it. The most common scaling failure happens when companies jump from one hundred test prospects to thousands overnight. Their winning approach generated strong response rates during testing, but the full-scale launch generates half those numbers because email deliverability tanks under the volume increase. They sent too many emails too quickly from domains that weren't warmed up for that volume. ISPs flag the sending patterns as suspicious, and emails start landing in spam folders instead of inboxes. It takes weeks to rehabilitate sender reputation, and during that time, momentum from the successful test evaporates.

Controlled scale prevents this by gradually increasing volume while monitoring deliverability health at every step.

Phase 2: Full Scale

Launch to your full database of qualified prospects while maintaining the systematic approach that made testing successful. Segment campaigns by persona with tailored versions for CROs, Directors, and Operations roles. Maintain weekly enrichment refresh using the SOPs from Chapter 9. Follow documented processes religiously rather than making ad-hoc changes.

Full scale doesn't mean launching to everyone simultaneously. It means expanding systematically to your entire addressable market over a controlled timeframe that preserves deliverability and message quality.

Multi-Channel Orchestration

Email serves as the foundation channel because it provides the highest volume potential and best tracking capabilities. But research increasingly shows that multi-channel approaches dramatically outperform single-channel outreach. Data from sales engagement platforms shows that multichannel prospecting is roughly three and a half times more effective than email-only campaigns. HubSpot's 2025 State of Sales Report found that social media now delivers forty-two percent response rates - nearly double email's twenty-six percent.

Add LinkedIn outreach selectively for high-value prospects where you have mutual connections, shared experiences, or strong engagement signals from their profile activity. Don't randomly add LinkedIn to every prospect - use enrichment data to identify where social engagement provides additional value.

Include phone outreach strategically for timing-signal prospects where funding, hiring, or leadership changes suggest immediate evaluation urgency. Phone works best when you can reference specific recent events that create conversational relevance.

The key is data-driven orchestration rather than random channel mixing. Use enrichment insights to determine the optimal channel mix for each prospect based on their situation, preferences, and timing. Verkada, the enterprise security company, took this approach to its logical conclusion by using Clay to automatically generate personalized landing pages for each prospect - six hundred unique pages at the click of a button, referencing job titles, company logos, and specific context. They use these pages across ABM campaigns, outbound sequences, inbound follow-ups, and even ad campaigns. That's what systematic multi-channel orchestration looks like when powered by enrichment data.

Campaign Monitoring and Rapid Adjustment

Daily monitoring prevents small problems from becoming campaign disasters. Track sends, opens, responses, and meetings through real-time dashboards that show trends and anomalies immediately.

Weekly reviews identify what's working and what's degrading before problems compound. Look for response rate changes, deliverability degradation, and message quality issues that require immediate adjustment.

Don't wait for monthly reviews to make changes - adjust in real-time based on data signals. If a subject line starts underperforming, switch traffic immediately. If response rates drop for a specific persona segment, investigate and adjust messaging quickly. The companies that sustain high campaign performance treat optimization the way Netflix treats recommendations: as a continuous process, not a quarterly review.

Common Campaign Failure Modes

Understanding how campaigns fail helps prevent expensive mistakes and wasted effort. These patterns repeat across dozens of implementations.

Failure Mode 1: "Perfect is the Enemy of Launched"

Companies that spend months perfecting campaigns before testing consistently underperform companies that test quickly and optimize based on real data. The company that launches an imperfect test in week two of development, learns from the failure, and iterates will beat the company that spends twelve weeks crafting the perfect campaign every single time. The first company has real market data. The second has assumptions.

The fix is disciplined adherence to test-first methodology. Resist the urge to perfect campaigns before testing. Launch imperfect tests quickly, learn from real market feedback, and optimize based on data rather than assumptions.

Failure Mode 2: "Over-Automation Trap"

Sending campaigns at scale without human oversight creates embarrassing mistakes that damage credibility and relationships. AI-generated personalization hooks that aren't reviewed by humans before sending will eventually congratulate a CEO on funding that was announced eighteen months earlier, reference a leadership change that never occurred, or mention an acquisition that was merely a rumor.

The worst version of this I've seen involved congratulating a company on their "recent Series B" when the company had actually laid off thirty percent of their workforce the previous month. The CEO forwarded the email to his network with the comment: "This is why I don't trust vendors who claim to do research."

The fix is implementing quality check SOPs from Chapter 9 that ensure human review of AI-generated content before sending. Automation should amplify human intelligence, not replace it entirely.

Failure Mode 3: "Data Rich, Insight Poor"

Having comprehensive enrichment data but using it generically wastes your investment and produces results no better than basic outreach. The classic mistake looks like this: "Hi John, I noticed your company has 250 employees and uses Salesforce. Many companies with 250 employees that use Salesforce struggle with efficiency challenges. Our solution helps companies like yours improve efficiency."

This messaging uses enrichment data but doesn't create genuine personalization or relevance. It's template language with data insertion rather than contextual insight.

Better usage of the same data sounds like this: "Hi John, I noticed LogisticsCorp is using Salesforce but doesn't have any automation tools connected to it yet. Most logistics companies at your scale run into the same challenge: as you grow past two hundred fifty employees, manual data entry becomes a real bottleneck for the sales team. We've helped similar companies automate their workflow while maintaining the relationship focus that wins deals in logistics. Worth discussing?"

The fix is implementing SOPs for enrichment-to-personalization conversion that ensure data insights translate into contextual, relevant messaging rather than generic templates with data insertion.

Failure Mode 4: "Scale Without Foundation"

Companies that skip testing phases and launch to entire databases can't isolate what works when results are mixed. They send three different approaches to thousands of prospects simultaneously with no control group, get mediocre overall results, and can't determine which approach was working because they mixed all variables. They have to pause everything, rebuild systematically, and restart with proper testing methodology - losing months of pipeline generation in the process.

The fix is disciplined adherence to the two-week testing framework before any scale activities, regardless of pressure to launch quickly.

Failure Mode 5: "Set It and Forget It"

Launching campaigns without ongoing monitoring and optimization leads to gradual performance degradation. Every campaign has a half-life. Strong response rates in month one become mediocre by month three as messaging becomes stale, subject lines become recognizable, and minor technical issues accumulate. Without daily monitoring and weekly optimization, companies don't notice the degradation until quarterly pipeline reviews reveal dramatically reduced meeting generation. By then, months of poor campaign performance have been wasted.

The fix is implementing daily monitoring dashboards and weekly review processes that identify performance changes immediately.

Campaign Architecture for Different Scenarios

Different business situations require different campaign approaches, timelines, and success metrics. The two-week testing framework adapts to various scenarios while maintaining systematic methodology.

New Market Entry requires extended testing phases. When expanding to new verticals or geographies, you can't assume that messaging which works in one market will translate to another. Healthcare buyers care about regulatory compliance and patient impact. Banking buyers focus on operational efficiency and cost reduction. A messaging approach that generates twenty percent response rates in banking might completely fail in healthcare. Plan for four weeks of testing across multiple messaging approaches, followed by controlled scale, then systematic rollout. Shift enrichment focus to industry-specific signals like regulatory changes, technology adoption patterns, and funding sources unique to the new market.

Product Launch campaigns emphasize education and relationship leverage rather than new prospect acquisition. When introducing new capabilities to existing markets, target your existing customer base and warm prospects who are already familiar with your core offering. The approach should progress from announcement to education and trial to full adoption. Concentrate enrichment on technology stack compatibility and usage patterns that indicate readiness for your new capabilities.

Competitive Displacement campaigns require patience and timing. When targeting prospects using competitor solutions, focus on contract renewal timing, dissatisfaction signals from review sites, and feature gap exploitation. This is an ongoing prospect-by-prospect approach rather than a mass campaign, driven by enrichment data that emphasizes technology change signals and competitive intelligence.

Account-Based Campaigns for high-value targets require multi-threading, coordinated engagement, and extended timelines. For your top twenty-five accounts, plan ninety-day campaigns with multiple stakeholder engagement across email, LinkedIn, phone, and direct mail. Each account gets deep enrichment across all five levels from Chapter 7, with personalized research on multiple stakeholders and customized value propositions for different buyer personas.

Reactivation Campaigns re-engage cold prospects by acknowledging the previous relationship while presenting new information that justifies renewed contact. Focus enrichment on what changed since the last interaction - new leadership, funding events, technology adoptions, or market developments that create fresh relevance. Sendoso used exactly this approach with their "Wake the Dead" play, automating the process of identifying dormant leads and crafting personalized outreach that acknowledged the previous relationship while presenting new context. The result was a significant portion of their million-dollar pipeline coming from prospects most companies would have written off.

What Full Execution Looks Like: A Three-Month Campaign Arc

Let me walk you through what systematic campaign execution looks like when the test-first methodology is applied with discipline.

Month one is testing and initial scale. Weeks one and two are pure testing: one hundred prospects divided across three approaches. The results will be clear. One approach bombs with low single-digit response rates. One is decent with moderate numbers. One is clearly winning with double-digit response rates and meetings booking. Weeks three and four are controlled scale: expand to five hundred prospects using the winning approach while monitoring that response rates hold within two to three percentage points of test results.

A strong month one generates forty-plus meetings from six hundred prospects contacted, creates six figures in pipeline, and establishes a cost per meeting under one hundred fifty dollars including enrichment, tools, and labor.

Month two is full scale and optimization. Launch to your full qualified database using segmented versions of the winning approach. You'll typically see higher response rates from senior executives, moderate rates from directors, and slightly lower but still strong rates from operations roles. Run ongoing optimization - subject line testing, send time adjustments, personalization hook refinement based on response patterns. Weekly enrichment refresh identifies new high-priority prospects with recent timing signals.

A strong month two generates significantly more meetings through scale efficiencies, creates seven figures in pipeline, and pushes cost per meeting down as volume increases.

Month three is systematic operation. The transition from "campaign launch" to "revenue system" happens here. Operations follow SOPs from Chapter 9. Weekly enrichment refresh identifies timing signals for opportunistic outreach. Continuous optimization maintains effectiveness. Team productivity reaches steady state with predictable daily activities, weekly optimization cycles, and systematic pipeline generation rather than sporadic launches.

By the end of ninety days, the best implementations generate two hundred-plus meetings, create multiple millions in pipeline, achieve meeting-to-opportunity conversion rates above thirty percent, and establish cost per meeting below one hundred twenty-five dollars - representing a seventy percent or greater reduction from traditional outbound methods.

The board presentation writes itself: systematic methodology that transforms prospect intelligence into pipeline generation through disciplined testing, measurement, and optimization. Most boards approve expansion and increased investment because the methodology produces measurable, attributable, and repeatable results.

Executive Takeaways

For the CEO: Campaign success requires infrastructure plus strategy plus execution working together systematically. The two-week testing framework prevents expensive mistakes by validating approaches before full-scale investment. Expected timeline is two weeks for testing, two weeks for controlled scale, then systematic full deployment. ROI typically ranges ten to twenty times the investment in the first ninety days when methodology is followed. The key insight is that campaign success isn't about creative brilliance - it's about systematic optimization based on real market feedback.

For the CRO: Testing beats planning every single time because market reality trumps theoretical strategy. Use enrichment data to determine campaign approach rather than guessing about prospect priorities and timing. Multi-channel orchestration should be based on data insights about prospect preferences rather than random channel mixing - the data shows multichannel approaches outperform single-channel by three and a half times. Monitor campaign performance daily and adjust weekly rather than waiting for monthly reviews.

For the CMO: Campaign architecture must match business scenario - new market entry requires different approaches than product launch or competitive displacement. Personalization at scale requires enrichment plus automation plus systematic processes working together seamlessly. Quality metrics matter more than volume metrics because relevant engagement converts better than broad reach. The Backlinko analysis of twelve million emails confirmed what we see in practice: personalized messages get thirty-three percent more replies.

For the CFO: Test-first methodology reduces both risk and cost by preventing expensive failures before they scale. Cost per meeting should drop sixty to seventy percent compared to traditional outbound methods through better targeting and messaging relevance. Campaign ROI is directly measurable and attributable when systematic tracking is implemented properly. Budget expectations: five to fifteen thousand dollars for test phases, then scaling investment based on proven results rather than hopeful projections.

For the CTO/RevOps: Campaign monitoring requires real-time dashboards that surface performance trends and anomalies immediately rather than retrospective reporting. Integration between Clay enrichment, your CRM, and your email automation must be bulletproof because campaign success depends on seamless data flow. Daily health checks prevent silent failures that destroy campaign effectiveness without obvious symptoms. Maintaining the infrastructure built in Chapters 7-9 is prerequisite for campaign success.

The transformation from random marketing activities to systematic campaign execution represents one of the most significant competitive advantages available to B2B companies today. While competitors continue hoping their outreach will generate pipeline, you'll be systematically creating it through data-driven methodology that improves continuously.

In Chapter 11, we'll show you how to optimize and sustain campaign performance over time - the testing frameworks, dashboards, and continuous improvement processes that prevent the inevitable decay and keep your campaigns performing at peak effectiveness month after month.

CHAPTER 11

Optimizing Campaign Performance with AI

Other than Tommy Boy, Top Gun, and Caddyshack, name something that's just as good on repeat as it was the first time you saw it.

You can't. Because everything else has a shelf life - including your best-performing sales campaign.

Netflix doesn't show you the same recommendations for six months. Amazon doesn't keep the same homepage. Google doesn't stop adjusting search results. Spotify doesn't play the same Discover Weekly playlist for a year. These companies have built the most sophisticated revenue engines on the planet, and not one of them has ever found a "winning formula" and stopped testing.

So why do sales teams find a campaign that's working and refuse to touch it?

Because it feels like a risk. When response rates are at 16% and meetings are flowing, the instinct is to protect what's working. Don't mess with it. Don't change the subject lines. Don't adjust the targeting. Don't tweak the sequences. Why fix what isn't broken?

Here's why: it's already breaking. You just can't see it yet.

Every campaign has a half-life. The email that got a 16% response rate in January will get 11% by March and 7% by June - even if you don't change a single word. Not because the email got worse. Because the market moved. Competitors saw your approach and copied it. Prospects who responded early are out of the funnel, and the ones left have seen similar messaging from three other companies. Your "personalized" openers are referencing trigger events that are now six months old. The data drifted while you were celebrating.

The decay is invisible until it's a crisis. One month you're hitting every target. Three months later, your pipeline has dropped 60% and nobody can explain what changed. Nothing changed on your end - that's exactly the problem. The market changed around you while your campaigns stood still.

The companies that sustain results - the ones that are still accelerating a year after launch - treat optimization the way Netflix treats recommendations: as a continuous process, not a quarterly review. They're testing new approaches every two weeks, refreshing enrichment data monthly, rotating messaging before fatigue sets in, and measuring leading indicators that tell them when decay is starting, not after it's already destroyed their pipeline.

This chapter is about building that optimization engine - the KPIs that actually predict performance, the testing frameworks that keep campaigns fresh, and the dashboards that tell you when something's slipping before it shows up in your revenue numbers.

The A/B Testing Engine

Building systematic testing infrastructure requires clear methodology for what to test, how to structure tests, and when to implement results. Random testing creates confusion rather than insights.

Building Your Testing Infrastructure

What to test should be prioritized based on potential impact on campaign performance. Subject lines affect forty to sixty percent of open rate performance. Opening lines affect thirty to fifty percent of response rate performance. Call-to-action language affects twenty to forty percent of meeting booking performance. Test elements in order of potential impact to maximize the value of your optimization time and effort.

How to structure tests requires control versus variant methodology that isolates specific variables while maintaining statistical validity. Control groups should represent your current best practice, while variant groups should test specific hypotheses about improvements. Sample size requirements determine statistical significance - you need at least one hundred prospects per test variation and at least fifteen responses per variation to achieve statistical confidence.

Testing velocity must balance learning speed with database preservation. The companies that win at optimization aren't the ones running massive quarterly tests - they're the ones running disciplined weekly tests. Analyze previous week's results on Monday. Design new tests Tuesday. Launch to two-hundred-prospect cohorts Wednesday. Monitor early indicators Thursday and Friday. Automate data collection over the weekend. Implement winners and design next tests the following Monday.

That rhythm produces fifty-two optimization cycles per year instead of four. Each test is smaller and faster, but the cumulative learning is dramatically higher than companies running big, slow tests quarterly. You don't need a massive data science team to do this - you need a disciplined process and the commitment to follow it.

Subject Line Optimization

Subject lines determine whether prospects open your emails, making them the most critical optimization element for email-based campaigns. Small improvements in open rates create large improvements in total campaign performance.

Testing framework should evaluate different psychological approaches: curiosity-based subject lines like "Interesting observation about Company," value-based subject lines like "3 insights for Industry teams," personalization-based subject lines like "Name, quick question," and urgency-based subject lines like "Brief question about Recent Event."

Systematic testing across our client implementations has consistently revealed that enrichment-specific subject lines outperform generic ones by forty to fifty percent. A generic subject line like "Quick question, First Name" typically generates twelve to fourteen percent open rates. An enrichment-specific subject line like "Thoughts on Company's recent expansion" generates twenty to twenty-four percent open rates using recent hiring or funding signals from Clay enrichment data.

The difference isn't just statistical - it's strategic. Enrichment-specific subject lines demonstrate immediate relevance and knowledge of the prospect's situation, creating curiosity about what insights the sender might have.

Another significant discovery: personalized subject lines with names perform differently based on seniority level. "Name, interesting observation" works well for Directors and VPs but performs poorly for C-level executives, who seem to find the casual tone inappropriate for their level. C-level executives respond better to company-focused subject lines like "Thoughts on Company Name's growth trajectory" that demonstrate business-level thinking rather than personal familiarity.

Opening Line Optimization

Opening lines determine whether prospects respond to your emails, making them the second most critical optimization element for campaign success.

Testing AI-generated versus human-written personalization reveals context-dependent results. In industries where enrichment data translates directly into business insights, AI-generated hooks from Clay often outperform human-written versions because they're more consistent, more relevant to enrichment data, and less likely to contain factual errors. The AI hooks try to be relevant while humans often try to be clever - and relevance beats cleverness every time in B2B outreach.

However, in industries requiring highly technical knowledge - financial services, certain areas of cybersecurity, specialized manufacturing - human-written hooks outperform AI because the industry requires nuanced understanding that AI can't yet replicate accurately.

The key insight is that context matters. Test both approaches for your specific market before committing to either. Research-based approaches that reference specific company actions consistently outperform insight-based approaches in response rates, but insight-based approaches attract more strategic prospects who convert at higher rates. The best optimization accounts for both volume and quality.

Call-to-Action Optimization

Call-to-action language significantly impacts meeting booking rates. Testing across multiple implementations reveals consistent persona-based differences.

Direct asks like "Worth a fifteen-minute call?" generate strong booking rates from CROs and senior executives who appreciate directness and have clear decision-making authority. The same direct ask generates much lower rates from Directors and mid-level managers who prefer softer approaches that let them control the pace of engagement.

Softer asks like "Should I send over some details?" generate lower rates from CROs but significantly higher rates from Directors. The psychology makes sense: CROs are comfortable making quick decisions, while Directors want to gather information before committing.

The most powerful finding across all testing: question-based CTAs consistently outperform statement-based CTAs across all personas and industries. "Are you the right person to discuss this with?" generates twenty to twenty-five percent higher response rates than "I'd like to discuss this with you." Questions invite participation. Statements request compliance. B2B prospects are more comfortable engaging with questions because they maintain control over the interaction.

Send Time Optimization

Send time testing reveals persona-specific patterns that significantly improve campaign performance when implemented systematically.

CROs and senior executives respond better to emails sent between six and seven in the morning - they're reading email before their schedules fill with meetings. Directors respond better to emails sent in the early afternoon, using breaks between meetings to catch up on email. Operations roles respond better to emails sent between eight and nine in the morning before diving into technical work.

Implementing persona-based send time optimization typically improves overall response rates by twenty-five to thirty percent without changing any other campaign elements. However, over-optimizing send times down to fifteen-minute increments creates diminishing returns and operational complexity that hurts more than it helps. Optimize by hour, not by minute.

Modern email platforms like HubSpot now offer AI-powered send-time optimization that analyzes individual prospect engagement patterns to predict optimal send times for each specific person. The system discovers that even within persona categories, individual prospects have dramatically different optimal times based on personal email habits, industry schedules, and engagement behaviors. Companies using AI send-time optimization see twenty-five to thirty percent improvement in open rates compared to persona-based timing, with some individual prospects showing much larger improvements when emails arrive at their personal optimal time versus standard business hours.

Sequence Structure Optimization

Testing sequence length and timing reveals optimal structures that balance persistence with respect for prospect attention.

Systematic testing of three-touch, four-touch, and five-touch sequences consistently produces the same finding: four-touch sequences hit the sweet spot. Three-touch sequences leave potential responses on the table. Four-touch sequences generate the highest total response rate with optimal balance between persistence and respect. Five-touch sequences actually reduce total response rates because the fifth touch creates fatigue.

Gap timing between touches matters just as much as the number of touches. Three-day gaps feel pushy and reduce response rates. Seven-day gaps lose momentum and prospects forget about you. Five-day gaps provide optimal balance - enough time for prospects to process information while maintaining enough urgency to sustain engagement.

The compound effect of sequence optimization is significant: better spacing between touches improves response rates to individual emails while also improving overall sequence performance.

Data-Driven Performance Analysis

Building systematic performance analysis requires the right combination of real-time monitoring and periodic deep analysis. Too much data creates paralysis. Too little data creates blind spots.

Building Your Performance Dashboard

Dashboard design should focus on actionable insights rather than comprehensive data display. The temptation is always to track everything. Resist it. Companies that track forty-seven different metrics across eight different dashboards create information overload that paralyzes decision-making rather than enabling it. More data doesn't create better decisions - better data creates better decisions.

The simplified approach focuses on three dashboards. The first is Operational Health, tracking daily technical metrics: sends completed, deliverability rates, bounce rates, spam complaints, and open rates. These metrics identify technical problems requiring immediate attention. The second is Performance Trends, tracking weekly optimization metrics: response rates, meeting booking rates, conversion quality, and competitive intelligence. These identify optimization opportunities and performance degradation before they impact pipeline. The third is Business Impact, tracking monthly ROI metrics: pipeline generated, cost per meeting, ROI attribution, and strategic insights. These demonstrate business value and inform strategic decisions.

Leading vs. Lagging Indicators

Understanding the relationship between leading and lagging indicators enables predictive optimization rather than reactive problem-solving.

Leading indicators predict future performance: open rates, response rates, meeting booking rates, and email deliverability scores. These metrics change quickly and provide early warning signals. Lagging indicators measure past performance: pipeline generated, deals closed, revenue attributed, and customer acquisition costs. These change slowly but reflect ultimate business impact.

Both matter for different reasons. Leading indicators enable rapid optimization and problem prevention. Lagging indicators measure business value and strategic success.

The common mistake is optimizing exclusively for leading indicators without monitoring lagging indicators. You can improve open rates and response rates dramatically, but if the optimization attracts unqualified prospects who respond but don't convert to opportunities, you've optimized for engagement rather than business outcomes. Higher response rates don't matter if the responses come from prospects who can't buy.

The fix requires rebalancing optimization to consider both. Test subject lines and opening hooks for response rate improvement while also tracking meeting quality and opportunity conversion to ensure optimization doesn't sacrifice business outcomes for engagement metrics.

Cohort Analysis for Campaign Optimization

Comparing performance across different prospect segments reveals optimization opportunities that aggregate analysis misses.

The most common discovery from cohort analysis is that overall response rates mask dramatic differences between segments. A fifteen percent overall response rate might hide the fact that your fifty-to-two-hundred-employee segment generates twenty-four percent response rates while your five-hundred-plus segment generates eight percent. This insight completely changes your ICP definition and campaign targeting. Instead of targeting "mid-market companies," you refine focus to "growing companies with fifty to two hundred employees" and see overall response rates improve to twenty-plus percent while meeting quality increases dramatically.

You stop trying to be everything to everyone and start being exactly right for someone. Cohort analysis shows you who that someone is.

Enrichment Quality Impact Analysis

Testing campaign performance with different levels of enrichment reveals clear ROI relationships between data investment and campaign results.

The difference between standard enrichment - firmographic, technographic, and contact data - and deep enrichment that adds intent signals and personalization data is dramatic. Standard enrichment typically supports sixteen percent response rates. Deep enrichment supports thirty-plus percent response rates - more than double - while actually reducing cost per meeting because the additional meetings generated far outweigh the extra enrichment cost.

The math is compelling. Spending an extra four dollars per prospect on deeper enrichment generates enough additional meetings to produce over one hundred seventy times return on the enrichment investment. Deep enrichment isn't expensive - it's profitable. Every dollar spent on better data generates multiples in additional pipeline value.

This is why the waterfall enrichment approach from Chapter 7 matters so much. OpenAI doubled their enrichment coverage from low forty percent to high eighty percent using Clay's multi-provider waterfall approach. Anthropic tripled their enrichment rates. That additional coverage translates directly into better campaign personalization and higher response rates.

Message Resonance Analysis

Understanding which personalization approaches generate the highest engagement reveals insights that inform future campaign development.

Analysis across multiple implementations reveals consistent patterns. Recent funding mentions generate thirty-four percent higher response rates than technology stack references, but technology stack references generate higher meeting quality scores. Hiring signals outperform funding signals by twelve percent in response rates and twenty-three percent in meeting booking rates because hiring signals indicate immediate operational challenges rather than general growth. Company-specific insights beat industry-level insights by twenty-three percent because they demonstrate individual attention rather than broadcast knowledge.

Personal connections - mutual LinkedIn connections, shared experiences - generate the highest response rates by far, averaging forty-seven percent. But they're only applicable to about twelve percent of prospects, limiting scalability.

The optimization priority becomes clear: use personal connections when available, prioritize hiring signals over other timing signals, reference company-specific insights over industry trends, and use funding information for context rather than primary hooks.

AI-Powered Optimization Techniques

Artificial intelligence enables optimization approaches that would be impossible with manual methods, creating competitive advantages through automated testing, predictive insights, and personalized optimization.

Predictive Send-Time Optimization

Instead of optimizing send times by persona, AI can analyze individual prospect engagement patterns to predict optimal send times for each specific person. The system discovers that even within persona categories, individual prospects have dramatically different optimal send times based on personal email habits, industry schedules, and engagement behaviors. AI doesn't just optimize for averages - it optimizes for individuals. Instead of sending to all CROs at seven in the morning, you send to each CRO at their personal optimal time based on historical engagement patterns.

AI-Powered Subject Line Generation

Using GPT and other AI tools to automatically generate subject line variations enables continuous optimization at scale without manual creative bottlenecks. The process works like this: AI generates ten subject line variations based on enrichment data and performance patterns, automatic A/B testing runs across small cohorts, the winning variation scales to the full campaign, and results feed back into the AI for better future generation.

Companies implementing this approach see continuous improvement in open rates - typically from the low teens to the high teens or low twenties over six months - with AI-generated subject lines consistently outperforming human-created alternatives. The AI doesn't get tired, doesn't run out of ideas, and doesn't have creative blocks. It continuously generates variations, tests them, and gets better based on what works.

Sentiment Analysis for Response Optimization

AI-powered sentiment analysis of prospect responses enables automatic adjustment of follow-up messaging based on engagement quality and emotional tone.

Categorizing responses into positive, neutral, negative, and confused categories, then triggering different follow-up sequences based on sentiment classification, reveals a powerful insight: confused responses are often the most valuable leads.

Traditional approaches treat confused responses as negative. But prospects who ask clarifying questions are often more engaged than prospects who immediately say yes without understanding what you do. Creating separate educational tracks for confused responders typically improves conversion rates by forty-plus percent compared to treating confusion as rejection. Confusion isn't rejection - it's opportunity.

Predictive Response Scoring

AI can analyze prospect characteristics and engagement signals to predict response likelihood before outreach, enabling sales teams to prioritize follow-up efforts on prospects most likely to engage. The system identifies that prospects with recent funding plus hiring signals plus technology stack gaps have dramatically higher response probability than prospects with only firmographic matches.

The workflow becomes: focus immediate follow-up on high-probability prospects, standard follow-up for medium-probability, automated nurture for low-probability. Same team effort, dramatically better results because you're focusing energy where it matters most. Companies implementing predictive scoring typically see two to two-and-a-half times improvement in meeting booking rate with the same sales team effort level.

Automated Sequence Optimization

AI can automatically adjust sequence structure and content based on individual prospect engagement patterns. If a prospect opens emails but doesn't respond, extend the sequence with additional value-add content. If they don't open emails, try different subject line approaches and send times. If they engage with content but don't book meetings, add social proof and urgency elements. If they show high engagement, accelerate to direct meeting requests.

The result is that every prospect gets a unique campaign experience optimized for their specific engagement patterns, rather than everyone receiving the same static sequence. Companies implementing adaptive sequences see thirty-plus percent improvement in overall response rates.

Optimization Pitfalls to Avoid

Understanding common optimization mistakes prevents wasted effort and enables systematic improvement.

Over-Optimization happens when you test too many variables simultaneously. Running twenty-three different test variations at once makes it impossible to determine which variables drive performance. When overall results improve, you can't identify which optimizations to implement permanently. The fix: test one variable at a time with clear control groups and adequate sample sizes.

Under-Optimization happens when you find a winning campaign and stop testing. Every campaign has a half-life. Strong response rates in month one decline by month three and crater by month six due to market saturation, competitive mimicry, and message fatigue. The competitor who implements weekly optimization maintains strong response rates throughout the same period by continuously adapting. The fix: implement systematic weekly testing regardless of current performance.

Sample Size Errors happen when you declare winners based on too few responses. Fifteen total responses across test variations is not statistically significant. What looks like a winner at small scale may generate worse results than baseline at full scale. The fix: require at least one hundred prospects per variation and fifteen responses per variation before drawing conclusions.

Optimization Without Documentation creates a situation where you repeat the same tests because you can't remember what you've already tried. Six months of undocumented testing produces confusion rather than compounding knowledge. The fix: implement an optimization log that records every test, every result, and every implementation decision for future reference.

Ignoring Negative Results means you only implement winners without analyzing why other approaches failed. Understanding why something doesn't work is often more valuable than understanding why something does work, because it prevents future mistakes and refines your understanding of prospect preferences. The fix: document and analyze all results, not just the winners.

Building Your Optimization System

Systematic optimization requires infrastructure, processes, and discipline that transform ad-hoc testing into competitive advantage.

The Weekly Optimization Cycle

Monday morning: analyze previous week's performance data, identify trends and anomalies, document results from active tests, determine winning variations for implementation. Tuesday afternoon: design new tests based on previous week's learnings, create test variations and control groups, schedule launches, brief team on testing objectives. Wednesday morning: launch tests, confirm proper setup and tracking, monitor initial delivery. Thursday through Friday: monitor early results, identify technical issues, gather preliminary insights, prepare data for weekend collection. Weekend: automated data collection compiles engagement metrics for Monday analysis.

Running fifty-two optimization cycles per year compared to four transforms your competitive position. You improve faster than competitors can copy you. Weekly optimization becomes your competitive moat.

Creating Your Optimization Backlog

Systematic optimization requires a prioritized list of hypotheses to test, preventing random testing and ensuring efforts focus on highest-impact opportunities.

Backlog categories: message optimization including subject lines, opening hooks, and CTAs. Targeting optimization including persona segmentation, ICP refinement, and timing signals. Technical optimization including send times, sequence structure, and channel mix. Strategic optimization including offer testing, positioning variations, and competitive differentiation.

Prioritize by potential impact on key metrics, effort required for implementation, statistical requirements for valid testing, and strategic importance for business objectives. The backlog ensures systematic progression through optimization opportunities without running out of testing ideas or duplicating previous efforts.

The Optimization Meeting

Weekly optimization meetings should include RevOps for data analysis and technical implementation, sales leadership for front-line insights and team feedback, and marketing leadership for strategic direction and campaign coordination.

Meeting structure: review previous week's test results with statistical analysis, discuss implications and insights, decide which optimizations to implement immediately, design next week's tests, and assign ownership for setup and monitoring.

The transformation these meetings create is cultural as much as operational. They shift discussions from "I think this approach works better" to "the data shows this approach works better." When you have clear data about what works, debates become discussions about implementation rather than arguments about strategy.

Documentation and Knowledge Management

Recording test results in a centralized location builds institutional knowledge that compounds competitive advantage over time. Document the hypothesis, design, implementation details, results with statistical analysis, business impact, lessons learned, and recommendations for future testing.

Over eighteen months, a well-maintained optimization playbook capturing learnings across dozens of tests becomes worth millions in efficiency gains. New campaigns start from proven baselines rather than scratch. Repeated mistakes are avoided. Time to optimal performance accelerates. Competitors can copy your campaigns, but they can't copy your learning history.

Scaling Optimization Across Teams

Different teams running different campaigns can share optimization insights that improve performance across the entire organization. A subject line optimization from a new customer campaign might improve a prospect campaign by twenty-plus percent. A CTA optimization from an enterprise campaign might improve a mid-market campaign by thirty-plus percent.

The cross-pollination process: share optimization results across teams weekly, identify insights applicable to other campaigns, test successful approaches in different contexts, and document which learnings are universal versus context-specific. Optimization insights compound when they're shared systematically. Each team's learning benefits every other team's performance.

Executive Takeaways

For the CEO: Optimization creates sustainable competitive advantage that compounds over time because companies that learn faster build advantages that become increasingly difficult to overcome. Companies that optimize weekly consistently outperform companies that optimize quarterly by three to five times in campaign effectiveness. Expected improvement: fifty to one hundred percent performance gains over twelve months through systematic optimization. Required investment: five to ten hours per week across optimization activities. ROI calculation: every one percent improvement in response rate typically generates three to five percent improvement in pipeline.

For the CRO: Optimization is how good campaigns become great campaigns and how great campaigns stay ahead of competitive threats. Weekly testing rhythm prevents campaign stagnation and maintains competitive edge. Focus on leading indicators that predict future performance, not just lagging indicators that measure the past. Your competitive advantage comes from learning faster than competitors, not just performing better initially.

For the CMO: Campaign optimization requires systematic testing infrastructure that enables rapid learning without operational complexity. A/B testing at scale needs proper sample sizes and statistical rigor to generate reliable insights. Document everything because institutional knowledge compounds optimization value over time. Balance testing velocity with execution quality - test enough to improve continuously, not so much that you lose focus.

For the CFO: Optimization ROI is directly measurable and consistently significant. Typical improvements: fifty to one hundred percent response rate gains over twelve months. Cost structure: mostly time investment with minimal incremental technology costs. Payback period: usually within two to three weeks of implementing winning approaches. Long-term value: optimization capabilities become competitive moats that compound advantages over time.

For the CTO/RevOps: Optimization requires technical infrastructure including real-time dashboards, automated testing capabilities, and systematic documentation systems. Integration between Clay, your CRM, and your email automation must support rapid testing without creating operational complexity. Data quality determines optimization effectiveness - invest in measurement infrastructure before scaling testing activities.

The transformation from launched campaigns to optimized campaigns represents the difference between short-term tactical success and long-term strategic advantage. Campaign optimization isn't about perfect campaigns - it's about systematically improving campaigns that compound competitive advantages over time. The companies that optimize weekly don't just perform better - they create sustainable advantages that become increasingly difficult for competitors to overcome.

In Chapter 12, we'll show you how to measure and prove the ROI of everything you've built - the benchmarks, attribution models, and reporting frameworks that turn "it feels like it's working" into "here's exactly what it's worth."

CHAPTER 12

Measuring ROI & EBITDA Impact

There's a line that's been haunting marketers for over a hundred years. John Wanamaker, the merchant who essentially invented the department store, allegedly said it first: "Half the money I spend on advertising is wasted. The trouble is, I don't know which half."

It's been repeated so many times that it's practically a joke - the kind of thing someone says at a conference to get a knowing laugh before moving on. David Ogilvy quoted it. Every CMO has heard it. And for a century, everyone just accepted it as an unavoidable truth of doing business.

Here's what's remarkable: in the age of AI-powered sales, companies are STILL saying it. Not about advertising - about their entire revenue operation.

They can tell you meetings are up. They can tell you response rates are strong. They can show you dashboards full of pipeline metrics that all point in the right direction. But ask them the one question that actually matters - "How much did you spend and how much did you make?" - and the room goes quiet.

Meetings booked, response rates, pipeline coverage - these are the metrics sales teams celebrate. They're also the metrics that make board members' eyes glaze over. A board member doesn't care that your response rate went from 7% to 19%. They care about what that 19% turned into. How much revenue? At what cost? Over what time period? And can you prove it was the AI investment that drove it, or did you just have a good quarter?

If you can't answer those questions with specifics - dollar amounts, payback periods, attribution models that hold up to scrutiny - then you're running Wanamaker's equation all over again. Half your AI investment is probably working. You just can't prove which half. And in a boardroom where every dollar of spend needs to justify itself, "probably working" gets your budget cut.

The companies that protect their AI investment - the ones that get expanded budgets, board-level endorsement, and the green light to scale - are the ones that translate operational metrics into financial proof. Not "pipeline is up." Instead: "$127,000 invested, $3.8 million in EBITDA impact, six-week payback period, 30x return." That's a sentence a board member remembers. That's a sentence that gets your methodology replicated across an entire portfolio.

This chapter is about building that sentence for your company - the benchmarks, attribution models, and reporting frameworks that turn "it feels like it's working" into "here's exactly what it's worth."

Executive Takeaways

For the CEO: ROI measurement isn't optional - it's how you prove business value and secure continued investment in AI capabilities that create sustainable competitive advantages. Clear ROI stories enable confident board presentations, investor communications, and strategic decision-making based on proven business impact rather than operational assumptions. Expected ROI range for well-implemented AI Sales Engines typically achieves 20-50x returns in Year 1, with 30x being conservative baseline expectation. Time to positive ROI typically ranges from 6-12 weeks for initial pipeline impact, with full payback occurring within 8-16 weeks depending on sales cycle length. Strategic value extends beyond immediate ROI through proven systematic growth capabilities that enable scaling AI across organization beyond sales function, creating compound competitive advantages.

For the CFO: AI Sales Engine ROI is measurable, attributable, and defensible with proper methodology that satisfies audit requirements and stakeholder scrutiny. Investment structure typically allocates 60% to tools and technology, 25% to implementation services and training, and 15% to ongoing optimization and support activities. Payback period typically occurs within 2-4 months of initial investment, with ongoing ROI maintaining 15-25x returns through continuous optimization and expansion. EBITDA impact creates 2-3x improvement versus revenue-only growth because AI combines revenue increases with cost reductions and operational efficiency gains. Valuation implications include 15-30% premium multiples for systematic growth demonstration versus founder-dependent or market-dependent growth patterns. Recommend conservative attribution methodology that maintains credibility with board and investor audiences while providing defendable ROI calculations.

For the CRO: Your ability to prove ROI determines budget allocation and team resources for continued AI investment and expansion across revenue organization. Focus on business outcomes including pipeline generation, deal closure rates, and revenue attribution rather than activity metrics that don't correlate to business value. Attribution methodology should use multi-touch analysis that fairly credits AI's influence across entire sales process while maintaining conservative assumptions. Forecast confidence typically improves 40-60% through predictive analytics and systematic pipeline management, creating strategic value beyond immediate revenue impact. Present ROI quarterly to demonstrate consistent value creation and justify continued optimization investment and strategic expansion initiatives.

For the CMO: Marketing attribution improves dramatically with AI Sales Engine implementation because every touchpoint is tracked, measured, and connected to revenue outcomes through integrated systems. Cost per acquisition typically decreases 50-70% through AI optimization of targeting, messaging, and conversion processes while deal quality and size increase simultaneously. Marketing efficiency measured as pipeline per marketing dollar typically improves 2-3x through better targeting and personalization capabilities. ROI story enables marketing budget expansion when you can prove definitive returns on marketing activities and demonstrate clear connection between marketing investment and revenue generation. Focus on multi-touch attribution that shows coordinated marketing and sales value creation rather than competing for credit.

For the Board: AI Sales Engine ROI is measurably superior to traditional sales investments because it combines revenue growth with cost reduction and systematic scalability that creates sustainable competitive advantages. Typical returns range from 20-50x in Year 1, with ongoing optimization maintaining 15-25x returns in subsequent years through continuous improvement and market expansion. EBITDA impact delivers 50-100% improvement through combined revenue gains and efficiency improvements that flow directly to profitability and cash generation. Valuation implications include systematic versus heroic growth demonstration that creates 15-30% valuation premiums at exit through reduced risk and proven scalability. Strategic importance reflects AI capabilities becoming table stakes for competitive relevance, with companies lacking systematic AI sales capabilities facing growing disadvantage in market competition and strategic positioning.

The transformation from operational success to financial proof represents the difference between hoping AI investments are working and knowing they're generating measurable business value. Your enrichment engine from Chapter 7 provides measurable improvements in targeting accuracy and personalization effectiveness. Your CRM integration from Chapter 8 enables comprehensive tracking and attribution of revenue impact. Your optimization system from Chapter 11 generates compound improvements that increase ROI over time rather than degrading.

ROI measurement transforms AI from experimental technology to strategic business capability that enables confident scaling, continued investment, and competitive advantage development. Companies that measure ROI systematically don't just prove current value - they build capability for future value creation through data-driven optimization and strategic expansion.

In Chapter 13, we'll show you how to scale and implement your proven AI Sales Engine across new markets, products, and opportunities while maintaining the systematic approach that creates predictable business results and measurable competitive advantage.

Chapter 13:

Implementing the AI Sales Engine

Implementation is 20% technology and 80% people. Most companies approach it like it's 80% technology and 20% people. That's why most implementations fail.

Read that again, because it's the single most important sentence in this book.

You can have the perfect ICP Blueprint. You can have three layers of intelligence feeding an enrichment engine that surfaces the exact right prospects at the exact right time. You can have a CRM integration that flows data seamlessly, SOPs that document every workflow, and a campaign framework that's been tested and validated. All of it - every chapter you've read to this point - can be flawless on paper and still fail completely in practice.

Not because the strategy is wrong. Because someone forgot that the strategy has to be executed by human beings who have habits, fears, quotas, egos, comfort zones, and a very limited appetite for change in the middle of a quarter.

Think about what you're actually asking a sales team to do. You're asking them to abandon the processes they've used - maybe successfully - for years. You're asking them to trust data from a system they've never seen instead of the gut instincts they've built over a career. You're asking them to learn new tools, follow new workflows, and change the way they start every single morning - all while maintaining the same quota, the same targets, and the same pressure to close deals this month.

That's not a software deployment. That's an organizational transformation. And the skills required to manage an organizational transformation are completely different from the skills required to configure a CRM or build an enrichment workflow.

Technology deployment asks: Is the system configured correctly? Are the integrations working? Is data flowing where it needs to go?

People management asks: Does the team understand why this change is happening? Do they believe it will make them more successful, not less? Have they seen proof that it works before being asked to adopt it? Is there a path to learn gradually instead of being thrown into the deep end? Are the early wins being celebrated so momentum builds instead of stalls? Is there a champion on the team - someone they respect - who's already using the system and getting results?

Most implementation plans answer every technology question and almost none of the people questions. They have detailed Gantt charts for data migration, integration testing, and workflow configuration. They have nothing for "how do we get the top rep who's been doing things his way for twelve years to actually use this?" They schedule a two-hour training session on a Tuesday afternoon, call it "enablement," and wonder why adoption is at 30% by Friday.

The companies that nail implementation - the ones that get 90%+ adoption rates and see results within the first month - don't treat it as a technology project with a people component. They treat it as a people project with a technology component. They phase the rollout. They build internal champions before they mandate adoption. They prove value in small, visible wins before they ask anyone to change their entire workflow. They create feedback loops so the team feels heard instead of bulldozed.

This chapter is about getting the 80% right - the stakeholder management, the phased rollout, the change management playbook, and the 90-day implementation roadmap that accounts for the reality that your biggest obstacle isn't the technology. It's the person sitting in the chair who hasn't decided yet whether this is going to help them or make their life harder.

Why Implementation Fails Despite Great Strategy

The gap between strategic brilliance and execution success reveals why most AI Sales Engine implementations fail despite having sound methodology, proven ROI, and adequate budget allocation.

The Strategy-Execution Gap

Most companies confuse planning with implementing, treating execution as a simple matter of following predetermined steps rather than managing complex organizational change in real-time.

The pattern repeats across industries and company sizes. A leadership team develops what they consider a perfect ninety-day implementation plan. Every process documented, every training session scheduled, every integration requirement mapped. The plan looks flawless on paper.

Then Week 3 arrives and everything falls apart.

The enrichment data that looked perfect during planning turns out to be thirty percent inaccurate when the full sales team starts using it. The CRM configuration that worked flawlessly in testing breaks under the load of a dozen reps importing data simultaneously. The workflow automation that seemed intuitive to the RevOps team is incomprehensible to sales reps who weren't involved in the design process.

This isn't a technology failure. The technical implementation succeeds - all the tools work, all the integrations function, all the data flows correctly. But the organizational implementation is a disaster because the team focused on process execution rather than people management.

The data confirms this pattern at scale. Gartner's 2024 research found that only eleven percent of sales organizations are able to drive commercial success while executing a transformation. Eleven percent. That means eighty-nine percent of companies that attempt sales transformation fail to achieve both the transformation and commercial success simultaneously. And Forrester Research reports that nearly half of all CRM projects ultimately fail, with poor user adoption being the leading cause - not technical issues, not budget constraints, not strategic missteps.

The companies that succeed don't have better technology or bigger budgets. They treat implementation as a people project with a technology component, not a technology project with a people component.

The Four Implementation Failure Modes

Understanding the most common failure patterns helps companies avoid the predictable pitfalls that derail otherwise sound deployments.

Failure Mode 1: The Big Bang Approach. Companies try to implement enrichment, CRM integration, automation, and campaign optimization simultaneously, overwhelming teams with too much change too quickly. They roll out Clay enrichment to the entire database on Monday, launch the new CRM configuration on Tuesday, activate email automation on Wednesday, implement new SOPs on Thursday, and start A/B testing on Friday. By the following Monday, the sales team is in complete rebellion. Every single thing they used to do automatically now requires thinking, learning, and problem-solving. Instead of selling, they're spending all their time figuring out how to use new tools.

Gartner's Seller Skills Survey of over one thousand B2B sellers found that seventy percent reported being overwhelmed by the number of technologies required to do their work. That's their baseline level of tech overwhelm before you add new systems. Dumping everything at once on top of that existing overwhelm is a recipe for revolt.

The fix requires reverting to a layered approach: enrichment first, CRM integration second, automation third, optimization fourth. Each layer mastered before adding the next. Change capacity is limited - even positive change creates stress and cognitive load. Respect your team's ability to absorb and master new processes gradually.

Failure Mode 2: The Top-Down Mandate. Leadership decides to implement AI without involving the sales team in planning or decision-making, creating passive resistance that kills adoption regardless of technology quality. The CEO announces the implementation during an all-hands meeting with no consultation, no input on tool selection or process design, and no acknowledgment that this represents a significant change in how people do their jobs every day.

The passive resistance is immediate and devastating. Reps attend training but don't engage. They use new tools when managers are watching but revert when left alone. They comply technically while finding ways to minimize impact on existing workflows. Six months later, the tools are installed and the processes are documented, but actual behavior has barely changed. Adoption metrics look reasonable on paper, but business results show no improvement because the team is going through the motions without genuine engagement.

The fix requires collaborative planning where the sales team helps design the implementation approach, provides input on tool selection, and takes ownership of success metrics. Participation mandated, but enthusiasm cannot be.

Failure Mode 3: The Technical Obsession. RevOps teams build perfect technical systems while ignoring user experience, creating tools that work flawlessly but are too complicated for daily use. Every integration is perfect, every workflow automated, every data point tracked and analyzed. The system can enrich from seven different data sources, score using twelve different criteria, personalize using five different AI models, track across nine different touchpoints. It's technically magnificent and practically unusable.

The sales team takes one look and rejects it entirely. Too many fields, too many screens, too many decisions, too many things that can go wrong. Adoption drops to twenty-something percent after six months, and the whole system has to be rebuilt with user simplicity as the primary design criterion.

A simple system that everyone uses is infinitely more valuable than a sophisticated system that no one uses. Adoption beats perfection every time.

Failure Mode 4: The Training Gap. Companies provide initial training but no ongoing reinforcement, leading to skill degradation and gradual reversion to previous methods. They conduct comprehensive training during implementation - full-day workshops, hands-on practice, detailed documentation, follow-up Q&A sessions. Initial adoption looks promising. Ninety-six percent completion rate on training modules.

Three months later, skills have degraded by sixty percent. Reps use only the simplest features, ignore advanced capabilities, and revert to familiar shortcuts. Complex skills that aren't used daily are forgotten within weeks. Advanced features requiring multiple steps are abandoned for simpler alternatives.

Training isn't a vaccination - one comprehensive dose doesn't provide permanent immunity. Sustainable adoption requires sustainable skill development: weekly skill-building sessions, monthly refresher workshops, quarterly advanced training, and ongoing individual coaching.

The 90-Day Implementation Roadmap

Ninety days provides the optimal timeframe for comprehensive implementation. Shorter timeframes sacrifice thoroughness for speed, leading to adoption problems. Longer timeframes lose urgency while allowing resistance to solidify. The three-phase structure builds momentum through early wins, develops confidence through successful piloting, and scales effectively through systematic expansion.

Phase 1: Foundation (Days 1-30)

Week 1: Team Preparation and Alignment. The implementation kickoff sets tone and expectations for the entire process. Rather than announcing decisions already made, present the vision while addressing concerns and establishing shared success metrics.

The best kickoff meetings I've seen start with the CEO acknowledging the team's concerns directly: "I know some of you are worried that AI will replace you or make your jobs more complicated. I understand that concern, and I want to address it directly. Our objective isn't to replace anyone - it's to make everyone more effective at the parts of selling you actually enjoy."

Then involve the skeptics: "Sarah, you mentioned that you're concerned about data quality. Would you be willing to be our data quality validator during the first month? We need someone with your eye for detail to make sure we're getting accurate information." By the end of the meeting, former skeptics become stakeholders with specific roles in ensuring success.

Stakeholder mapping identifies three groups: champions who will evangelize adoption (typically twenty percent of the team), resisters who need individual attention, and fence-sitters who will follow the majority. Champions get early access and additional training to become internal experts. Resisters get one-on-one conversations to understand and address their specific concerns.

Week 2: Enrichment Engine Setup. Rather than enriching the entire database immediately, focus on testing enrichment quality with a small batch of one hundred prospects selected specifically for validation accuracy. The sales team reviews each enriched prospect to assess accuracy of company data, contact information, and timing signals. This serves two purposes: ensuring data quality and building team confidence.

The biggest mistake companies make here is enriching thousands of prospects immediately, only to discover that a significant percentage of the data is inaccurate or irrelevant. The sales team sees bad data and assumes the entire system is unreliable. It takes months to rebuild that trust. The corrected approach validates quality with a small sample, adjusts enrichment parameters based on feedback, and builds confidence before scaling.

Week 3: CRM Configuration. Setup begins with creating custom fields for enrichment data, but not populating them until the sales team validates that the fields make sense and provide value. Import testing uses the same one hundred prospects from Week 2 to ensure data flows correctly without formatting errors or performance issues.

Sales team validation focuses on practical questions: Is this information useful? Is it displayed clearly? Can I find it quickly during a conversation? Does it help me personalize outreach? Skip this validation and you risk importing thousands of records with incorrectly formatted data - phone numbers as text strings, dates in wrong format, descriptions truncated mid-sentence.

The technical integration works perfectly, but the practical usability is terrible, so reps ignore it entirely.

Week 4: SOP Documentation. Rather than creating SOPs in isolation, involve the sales team in documenting both current processes (baseline) and proposed new processes (target state). Current process documentation provides fallback options if new processes encounter problems. New process documentation reflects team input about practical challenges and preferred workflows.

Pilot testing with two or three early adopters (the champions identified in Week 1) validates that new SOPs work in practice rather than just theory. SOPs written without sales input result in processes that are technically correct but practically impossible. They assume unlimited time for data entry and research, written by people who don't actually make sales calls every day.

Phase 1 success metrics: enrichment accuracy above ninety percent as validated by sales team, CRM data quality score above eighty-five percent, team understanding assessment showing readiness for next phase, and early adopter enthusiasm indicating champion development.

Phase 2: Integration and Training (Days 31-60)

Week 5: Full CRM Integration. With data quality validated in Phase 1, scale enrichment to the full target database - typically one thousand to five thousand prospects depending on company size. Complete CRM setup implements all custom fields, workflows, and automation, but only after confirming that integration can handle full-scale data volume.

Don't assume that integration working with one hundred records means it will work with five thousand. Scale creates completely different technical challenges. Load testing, performance monitoring, and error handling must validate that infrastructure can support daily operational requirements.

Week 6: Comprehensive Team Training. Training effectiveness requires full-day workshops rather than one-hour webinars, hands-on practice with real prospects rather than theoretical examples, and role-playing exercises simulating actual sales conversations.

Research on sales onboarding confirms what we see in practice. Studies show that effective onboarding programs cut ramp time by fifty percent and decrease time-to-productivity by up to seventy percent. Every dollar invested in training generates approximately four dollars and fifty-three cents in return - a three hundred fifty-three percent ROI. But the format matters enormously. One-hour webinar training produces around thirty-five percent skill retention after two weeks. Full-day in-person workshops with hands-on practice produce nearly ninety percent retention. Webinars teach concepts. Workshops teach muscle memory and confidence.

Training structure includes system navigation, personalization techniques using data insights, process workflows for daily routines, and troubleshooting common problems. Q&A and troubleshooting sessions address specific concerns that general training doesn't cover.

Week 7: Pilot Campaign Launch. Rather than launching to the full database, implement a controlled pilot with two hundred prospects and two or three sales reps - the champions from Phase 1 - to test campaign effectiveness and identify optimization opportunities.

The pilot follows the SOPs from Week 4, uses enrichment data from Week 5, and applies training from Week 6. This integration test reveals practical challenges that weren't apparent during component testing. Daily monitoring tracks response rates, meeting bookings, deliverability issues, and team feedback.

Companies that skip pilot testing and launch immediately to their full database learn a painful lesson about deliverability. Poor configuration at scale causes domains to be flagged as spam, destroying sender reputation and requiring months of rehabilitation. What seemed like unnecessary delay becomes massive time savings by catching and fixing problems with limited downside risk.

Week 8: Pilot Results and Optimization. Analysis focuses on practical learnings: which enrichment data proved most useful, which personalization approaches generated best responses, which SOP steps created confusion, and which training topics need reinforcement.

SOP refinement based on pilot feedback ensures documented processes reflect actual best practices rather than theoretical ideals. Success story sharing with the full team builds excitement for Phase 3 rollout by demonstrating concrete results from their own colleagues.

Peer success drives adoption more effectively than any training or presentation. When colleagues succeed with the new system, excitement shifts from skeptical compliance to eager participation. Seeing someone you work with every day achieve real results is more convincing than any executive presentation.

Phase 2 success metrics: technical integration at one hundred percent uptime with under two percent error rate, training completion for one hundred percent of team with above eighty percent skill assessment scores, pilot performance above ten percent response rate and above fifteen percent meeting booking rate, and team confidence survey indicating readiness for full launch.

Phase 3: Full Activation (Days 61-90)

Week 9: Full Team Rollout. All reps begin using the complete system with access to the full database, but with intensive support during transition. Daily fifteen-minute huddles share wins, troubleshoot issues, and maintain momentum. Manager shadowing ensures reps follow SOPs correctly and identifies individual coaching needs before bad habits develop. Real-time support availability via Slack or phone enables immediate problem resolution.

Without support infrastructure, reps who encounter problems create workarounds that seem easier but are less effective. By the time you notice, bad habits are established and difficult to correct. Intensive support during Week 9 prevents this by making help immediately available when confusion arises.

Week 10: Campaign Scaling. Volume scales gradually from the pilot two hundred prospects to the full campaign of two thousand-plus while maintaining daily monitoring. Track response rates, deliverability scores, and meeting bookings to catch trends that might indicate technical problems, market saturation, or message fatigue.

Rapid adjustment protocols enable fixing issues within twenty-four hours rather than waiting for weekly reviews. Speed of response prevents small problems from becoming large ones. Public celebration of wins maintains energy during the scaling period.

Week 11: Optimization Begins. First A/B tests launch using the optimization framework from Chapter 11, starting with simple tests like subject lines and send times rather than complex experiments. Weekly optimization meetings establish the systematic testing culture that drives continuous improvement. Individual coaching for struggling reps addresses adoption challenges before they become performance problems.

The individual attention matters. Identifying a rep whose response rates are significantly below average and providing targeted coaching - perhaps they're using generic personalization instead of specific enrichment insights - can improve their results dramatically in just a couple of sessions. That individual improvement becomes a team success story rather than a failure example.

Week 12: ROI Measurement and Review. Ninety-day ROI calculation involves the entire team in understanding business impact. Results presentation to leadership and board demonstrates systematic success and justifies continued investment. Expansion opportunity identification reveals next growth areas: additional products, new markets, cross-functional implementation.

Next ninety-day cycle planning maintains momentum by establishing continuous improvement objectives rather than treating implementation as a completed project.

Phase 3 success metrics: team adoption above eighty-five percent daily active usage, performance improvement with response rates more than fifty percent above baseline, meeting booking targets met or exceeded, and team satisfaction survey showing positive sentiment.

Change Management and Team Adoption

Successful implementation requires understanding and managing the psychological and organizational factors that determine adoption success.

Understanding Resistance

People resist change because it creates uncertainty about their ability to succeed, their job security, and their status within the organization. Resistance isn't personal failure - it's predictable human response to disruption.

The most common form I encounter is identity-based resistance. Top performers often view their research and personalization skills as their "secret sauce." When AI appears to replicate those skills, they feel threatened: "If AI can do what I do, what makes me valuable? Why would the company need me?"

The breakthrough comes when they realize AI doesn't replace their capabilities - it amplifies them. Instead of spending three hours researching five prospects, they research fifty. Instead of generic personalization, they develop genuinely strategic approaches because they have richer information to work with. AI doesn't make great sellers obsolete. It makes them superhuman. They can do in one day what used to take a full week, and they can do it better.

Building the Coalition

Successful change requires champions who evangelize adoption from within the team rather than mandates imposed from leadership. Champions are typically twenty percent of the team - early adopters and opinion leaders.

Here's a counterintuitive insight: your highest-performing rep might not be your best champion. Other reps view top performers as naturally talented, so their success with the new system doesn't feel replicable. Instead, select a middle-performer who is respected for work ethic and collaboration. When that person achieves dramatic results using the AI Sales Engine, their success is viewed as accessible and achievable by everyone rather than reflecting unique talent.

Formalize the champion's role. Give them time to help teammates, share best practices, and troubleshoot problems. Peer coaching from a credible, relatable colleague is more effective than leadership direction because it comes from shared experience.

Addressing Resisters

Resistance requires understanding underlying concerns rather than dismissing or forcing compliance. The most vocal resister is often someone who feels excluded from the decision-making process: "Nobody asked what I thought. I've been here eight years, I know our customers better than anyone, and suddenly I'm being told a machine knows better."

The solution is making them part of the solution rather than a victim of it. Ask for their expertise in validating enrichment accuracy, refining personalization approaches, or designing workflows that account for edge cases only experienced reps would know about. Once they're contributing to the design instead of receiving the mandate, their entire attitude shifts. They go from your biggest opponent to one of your most valuable contributors.

Creating Quick Wins

Early success stories build momentum more effectively than comprehensive training or detailed explanation. Quick wins should be visible, specific, and personally meaningful.

The best quick wins involve someone using enrichment data to identify a timing signal - a prospect who just received funding, hired aggressively, or changed leadership - and converting it into an immediate meeting. "I sent the email on Tuesday and had a meeting scheduled by Thursday. The enrichment data gave me the perfect reason to reach out at exactly the right time." That story is more convincing than any training presentation because it demonstrates practical, immediate value that other reps can visualize achieving themselves.

Celebrate the win publicly. Share the specific enrichment insights used. Encourage other reps to look for similar opportunities in their own prospect lists. Success breeds success when it's visible and shared.

Continuous Reinforcement

Adoption requires ongoing reinforcement rather than one-time training events. Skills degrade, enthusiasm wanes, and competing priorities create drift without systematic reinforcement.

Think of adoption like physical fitness. You can't work out once and stay fit forever. Consistent, ongoing investment in skill development creates consistent, ongoing improvement. Weekly fifteen-minute skill sessions, monthly advanced training workshops, quarterly optimization challenges, and ongoing peer coaching relationships maintain adoption rates above ninety percent. Without them, even enthusiastic adopters gradually drift back to comfortable old habits.

Structured knowledge management tools save new employees nearly five hours per week during onboarding, leading to almost twelve percent weekly productivity gains. These same tools help existing employees maintain and deepen their skills over time.

The Communication Plan

Systematic communication throughout the ninety-day period maintains momentum, addresses concerns, and celebrates progress.

Weeks one through four focus on vision and reasoning: weekly updates explaining why the company is implementing AI and how it will benefit team members personally. Weeks five through eight emphasize progress and early wins: sharing technical milestones, training completions, and pilot results. Weeks nine through twelve celebrate results and expansion planning: highlighting team achievements, individual success stories, and next-phase opportunities.

The most effective communications I've seen come directly from the CEO - short video messages each Friday sharing behind-the-scenes insights about board discussions, competitive intelligence about market opportunities, and personal reflections about team progress. These create personal connection between leadership vision and daily execution. Team members feel like they're part of something important rather than just following new procedures.

Managing Executive Expectations

Ninety days to measurable results is the realistic timeline, not thirty days despite vendor promises or competitive pressure. Leading indicators like response rates and meeting bookings appear before lagging indicators like closed deals and revenue impact.

Transparent communication about challenges prevents loss of confidence when inevitable problems arise. Problems are normal and manageable when expected and addressed systematically. Hiding implementation struggles from the board makes problems appear more serious when they're eventually discovered. Transparent reporting maintains support and enables additional resources to solve problems.

Common Implementation Pitfalls

Underestimating Time Requirements. Implementation consistently takes two to three times longer than planned because companies underestimate time for training, adoption, and cultural change. Plan for technology deployment speed but people change pace. Realistic timelines with buffer prevent the perception of failure when implementation proceeds at natural pace.

Ignoring Data Quality. Poor data quality destroys team confidence in the entire system. Teams judge credibility based on accuracy of information they can personally validate. If the first prospect a rep calls has left the company, and the second prospect's email bounces, that rep stops trusting all the data and goes back to manual research. Quality-first implementation prioritizes accuracy over speed, validates with sales team input, and establishes ongoing hygiene processes.

Insufficient Training Investment. One training session cannot establish competency with complex new processes. Skills degrade by sixty percent within thirty days without reinforcement. Continuous training architecture - initial workshops plus weekly skill sessions plus monthly advanced training plus ongoing coaching - maintains and develops competency over time.

No Executive Accountability. Without consistent executive attention, implementation loses priority when competing demands arise. It becomes an operational project rather than a strategic initiative, competing poorly for attention and resources. Executive sponsorship with weekly accountability maintains strategic priority and ensures rapid problem resolution.

Perfectionism Paralysis. Waiting for perfect configuration prevents the learning that only occurs through real-world usage. You could spend six months refining configurations, testing edge cases, and optimizing workflows before allowing any sales team usage. By then, market conditions have changed and competitive opportunities have been missed. Launch at eighty percent completeness and optimize the remaining twenty percent based on actual usage feedback.

Skipping the Pilot. Full-scale launch without pilot testing prevents identification of problems while they can still be managed with limited impact. Deliverability problems not apparent during small-scale testing can cause domain reputation damage requiring months of rehabilitation. Controlled pilot testing with one to two hundred prospects identifies and fixes problems with minimal downside risk.

Forgetting to Celebrate. Teams lose momentum without recognition. Achieving every technical milestone and performance target means nothing if the human effort goes unnoticed. Celebration planning includes milestone recognition, individual achievement acknowledgment, and ongoing appreciation during challenging change periods.

Scaling Beyond Initial Implementation

Successful initial implementation creates the foundation for strategic expansion that multiplies competitive advantage.

When to Scale. Scaling requires evidence of systematic success rather than early optimism. Ready-to-scale indicators: sustained performance above ten percent response rate for two-plus consecutive weeks, team adoption above eighty percent, technical stability with minimal errors, and clear positive ROI. Don't scale after one strong week - wait for consistent results over four weeks to prevent scaling inconsistency. But don't wait too long either, because competitive advantages erode if expansion is delayed beyond market windows.

Geographic Expansion. Replicating success across new regions requires adapting enrichment data sources, personalizing messaging for cultural differences, and adjusting timing for local business patterns. The core methodology works globally, but execution details require local adaptation. Plan sixty to ninety days per new region.

Product Line Expansion. AI Sales Engine methodology adapts to additional products by developing product-specific ICPs, messaging, and buyer personas while leveraging shared enrichment and technical infrastructure. Different products target different decision-makers and require different qualification criteria, but share common prospecting infrastructure. Plan thirty to sixty days per new product line.

Team Scaling. New reps joining an established AI Sales Engine become productive dramatically faster than traditional onboarding. Research shows effective onboarding programs cut ramp time by fifty percent and improve win rates by fourteen percent. With comprehensive system training, buddy pairing with experienced users, and documented SOPs, new reps start with systematic processes and enriched data rather than developing research skills from scratch. They become productive in weeks rather than months.

Cross-Functional Expansion. AI capabilities extend beyond sales into marketing automation, customer success proactive outreach, and account management. Oyster, the global employment platform, saved forty hours per month per sales rep after implementing Clay-powered enrichment and intent-based outbound workflows - and those same enrichment capabilities benefit every revenue-generating function. Marketing uses the same enrichment for campaign targeting, customer success uses timing signals for proactive outreach, account management uses company intelligence for expansion opportunities. Cross-functional implementation typically generates three times the ROI of sales-only implementation because the enrichment investment serves multiple use cases.

Executive Takeaways

For the CEO: Implementation success requires personal involvement that can't be fully delegated. Your role includes weekly check-ins, executive sponsorship that maintains strategic priority, obstacle removal, and milestone celebration. Expected timeline is ninety days to full implementation and measurable results. Change management matters more than technology deployment - implementation is twenty percent technical and eighty percent people and process. Budget forty to sixty hours of leadership time across ninety days. The Gartner data is clear: only eleven percent of sales organizations succeed at transformation. The difference between the eleven percent and the eighty-nine percent is almost entirely about how they manage the people side.

For the CRO: Your role is champion and coach, not project manager. Daily involvement during the first thirty days addresses concerns and builds confidence. Weekly involvement during days thirty-one through ninety provides coaching and optimization. Address resistance immediately - skeptics poison morale if concerns aren't managed proactively. Celebrate wins publicly because peer success drives adoption more than mandates. Expect temporary productivity decline during weeks two through four, with performance returning to baseline by week five and exceeding it by weeks eight through ten.

For the CMO: Marketing and sales alignment is critical during implementation because enrichment data benefits both functions. Coordinate timing carefully - avoid launching major marketing campaigns during sales team training periods. Share optimization learnings between functions because successful approaches often apply across both. Marketing efficiency typically improves forty to sixty percent using the same enrichment data and targeting approaches.

For the CFO: Implementation requires upfront investment before ROI materializes, with typical sixty to ninety day lag between costs and measurable revenue impact. Budget comprehensively for training time and optimization, not just tool costs - people costs typically represent forty percent of total investment. Track leading indicators during implementation. Preliminary ROI calculations at day sixty based on pipeline generation, full ROI analysis at day one hundred twenty based on closed revenue. Salesforce research shows that eighty-four percent of sales teams using AI report increased sales, but only organizations that invest properly in implementation and adoption realize those gains.

For the CTO/RevOps: You own technical implementation but not organizational adoption - partner closely with sales leadership for change management. Prioritize user experience over technical perfection. An eighty percent ready system that everyone uses creates more value than a hundred percent perfect system that struggles with adoption. Daily monitoring during the first thirty days catches issues within twenty-four hours. Document everything because implementation learnings enable faster scaling. Plan for twenty percent buffer on all timelines because unexpected challenges always emerge.

The companies that nail implementation don't treat it as a technology project with a people component. They treat it as a people project with a technology component. They phase the rollout. They build champions before mandating adoption. They prove value in small, visible wins before asking anyone to change their entire workflow. They create feedback loops so the team feels heard instead of bulldozed.

Your enrichment engine from Chapter 7, your CRM integration from Chapter 8, your SOPs from Chapter 9, your campaign framework from Chapter 10, and your optimization system from Chapter 11 are all powerful tools. But tools don't create results. People following systematic processes create results. Implementation is how you bridge the gap between powerful tools and powerful outcomes.

In Chapter 14, we'll look ahead to where AI sales technology is heading and how to future-proof your investment so you're positioned to capture competitive advantages as the technology evolves.

Chapter 14:

Future-Proofing Your Revenue Engine

In 1975, a Kodak engineer named Steve Sasson built the first digital camera. It was the size of a toaster, took 23 seconds to capture an image, and stored it on a cassette tape. It was also the most important invention in the history of photography.

Sasson brought it to Kodak's leadership. They looked at it, understood exactly what it meant, and made a decision that would eventually destroy the company: they buried it. Kodak held the patent on digital photography - the technology that would transform their entire industry - and chose not to develop it because it threatened their film business. Film was profitable. Film was predictable. Film was working.

For the next twenty years, Kodak kept selling film while competitors built on the digital technology Kodak had invented. By the time Kodak finally tried to catch up, it was too late. They filed for bankruptcy in 2012, brought down not by a technology they didn't understand, but by a technology they invented and refused to evolve.

The parallel to AI sales is uncomfortable, and it should be.

Right now, you've built something powerful. The AI Sales Engine is generating results - pipeline is up, meetings are flowing, revenue is growing. Everything in this book has been leading to this point, and it's working. The natural instinct is to protect it. Don't change the enrichment workflows. Don't mess with the campaign sequences. Don't overhaul the SOPs. You found what works. Stay the course.

That instinct will kill your competitive advantage within eighteen months.

The AI Sales Engine you built today - the one generating those results right now - is already being matched by competitors who are implementing similar approaches. The enrichment tools you're using are available to everyone. The CRM integrations you configured aren't proprietary. The campaign frameworks that feel like your secret weapon are being taught at conferences and written about in blog posts. Your current advantage isn't permanent. It's a head start, and head starts have an expiration date.

Meanwhile, the next generation of capabilities is already emerging. Autonomous AI agents that respond to prospect inquiries in minutes, not hours. AI-generated personalized video at scale. Predictive models that tell you which deals will close before the prospect has even scheduled a demo. Companies building on these capabilities today will have them operational in twelve to eighteen months - right about the time your current approach starts feeling stale.

The question isn't whether your AI Sales Engine works. It does. The question is whether you'll do what Kodak did - hold the future in your hands and choose to protect the present - or whether you'll build the organizational muscle to continuously evolve, adopt, and stay ahead of a technology curve that never stops moving.

This chapter is about making sure you never become Kodak.

The Accelerating Pace of AI Evolution

AI Isn't Standing Still

The pace of AI development isn't linear - it's exponential. Each advancement creates the foundation for faster, more sophisticated advancements. What seems like cutting-edge technology today becomes table stakes tomorrow and obsolete the day after.

In 2023, AI sales tools meant basic automation and template personalization. Companies that implemented email automation and simple CRM integration considered themselves ahead of the curve.

In 2024, the bar moved to predictive analytics and sophisticated enrichment. Companies needed AI-powered lead scoring, intelligent sequencing, and data-driven personalization to remain competitive. Salesforce research showed forty-three percent of sales reps actively using AI, up from twenty-four percent the year before - a seventy-nine percent year-over-year increase. Eighty-one percent of sales teams were already experimenting with or had fully deployed AI tools.

In 2025, we saw autonomous agents and real-time adaptation. The companies winning had AI systems that operate independently, learn continuously, and adapt automatically to changing market conditions. Clay's Claygent AI research agent surpassed one billion cumulative runs, processing five hundred thousand research and outreach tasks per day - a scale of intelligence gathering that would have been unimaginable two years earlier.

In 2026 and beyond, we'll see capabilities that seem like science fiction today: AI that understands context and emotion as well as humans, autonomous systems that handle entire sales processes from prospecting to closing, and predictive intelligence that identifies opportunities before prospects know they have problems. Gartner predicts that forty percent of enterprise applications will feature task-specific AI agents by the end of 2026, up from less than five percent in 2025. Even more dramatically, they predict ninety percent of all B2B purchases will be handled by AI agents within three years, channeling more than fifteen trillion dollars in spending through automated exchanges.

The gap between leaders and laggards isn't narrowing - it's widening exponentially as compound advantages create insurmountable competitive moats.

What's Coming in the Next 12-24 Months

Voice AI for Sales Conversations. Real-time conversation coaching during sales calls isn't science fiction - it's happening now. AI systems listen to sales conversations, analyze prospect responses, and provide real-time coaching through the rep's screen. The AI doesn't replace human judgment - it augments human intuition with data-driven insights during live conversations. Automated note-taking and action item extraction eliminate fifteen to twenty minutes of administrative work after each call. Sentiment analysis during live conversations alerts reps when prospects are confused, concerned, or losing interest, enabling immediate course correction.

Autonomous SDR Agents. The biggest shift is from AI-assisted sales to AI-autonomous sales. Instead of helping humans execute, AI agents execute with human oversight. Current implementations show sixty percent of initial qualification happening without human intervention - the AI handles routine qualification questions, identifies buying authority, confirms budget availability, and determines timeline through natural conversation.

This isn't about replacing SDRs - it's about elevating them from administrative tasks to strategic activities. Human involvement shifts from execution to strategy and oversight. Instead of manually qualifying twenty prospects per day, reps review AI-qualified opportunities and focus entirely on relationship building and complex problem solving.

But there's a cautionary note. Gartner also predicts that over forty percent of agentic AI projects will be canceled by the end of 2027 due to escalating costs, unclear business value, or inadequate risk controls. And Forrester predicts that three out of four firms attempting to build advanced agentic architectures independently will fail due to complexity. The companies that succeed will be the ones that implement thoughtfully with clear ROI metrics rather than chasing capabilities for their own sake.

Predictive Deal Scoring. Deal scoring has evolved from basic lead scoring to sophisticated outcome prediction with eighty-five-plus percent accuracy. AI analyzes hundreds of variables - prospect behavior, company characteristics, interaction patterns, timing signals - to predict which opportunities are most likely to close. This transforms resource allocation from optimistic hope to mathematical probability.

Real-Time Competitive Intelligence. AI monitoring of competitor activities provides advantages impossible with manual research. AI agents track competitor pricing changes, messaging evolution, customer reviews, job postings, and market positioning shifts. Response time advantage: hours instead of weeks for competitive positioning adjustments.

Hyper-Personalized Video at Scale. AI-generated personalized videos represent the next evolution of personalization. Instead of personalizing text, AI creates custom video messages for each prospect using voice synthesis, image customization, and content adaptation. Early tests show three times response rates compared to text-only personalization because video creates emotional connection that text cannot match. The technology isn't perfect yet, but it's improving rapidly and will become mainstream within eighteen months.

The Competitive Advantage Timeline

Understanding adoption timing helps companies position strategically. Early adopters implementing capabilities while they're still in beta gain twelve to eighteen month advantages that compound through faster learning. Fast followers who adopt capabilities quickly after general availability maintain relevance without leading-edge risk. Laggards who wait for maturity face permanent disadvantage because early adopters' compound advantages become insurmountable.

Gartner's January 2025 poll of over three thousand respondents found that only nineteen percent of organizations had made significant investments in agentic AI, while forty-two percent made conservative investments and thirty-one percent were in wait-and-see mode. The gap between the nineteen percent and the thirty-one percent will be enormous within two years.

The lesson: competitive advantages from AI adoption are real but temporary. Maintaining them requires continuous evolution.

Building an Adaptable AI Architecture

Future-proofing requires technical architecture designed for evolution rather than optimization of current capabilities.

The Modular System Approach

Think of your AI Sales Engine like a smartphone rather than a landline telephone. Smartphones receive software updates adding new capabilities without hardware replacement. Individual apps upgrade or get replaced without affecting other functions. Traditional sales systems are like landlines - replacing any component requires replacing everything. Modern AI Sales Engines should be like smartphones - individual capabilities evolve while maintaining system integrity.

The Three-Layer Architecture

Layer 1: Data Foundation. Your enrichment infrastructure - Clay, ZoomInfo, industry databases - and CRM system represent the foundation that changes slowly and provides stability for rapidly evolving capabilities above. Choose platforms with robust APIs, comprehensive integration capabilities, and long-term viability rather than optimizing for current features. Updates typically needed every two to three years, driven by scale requirements rather than capability limitations.

Layer 2: AI Capabilities. This is the rapidly evolving layer where current capabilities like enrichment, personalization, and automation sit alongside emerging capabilities like autonomous agents and voice AI. This layer evolves every six to twelve months as new tools become available. Strategy: use best-in-class tools for specific functions rather than all-in-one platforms that compromise individual quality. Build modular connections so switching tools takes weeks rather than months.

Layer 3: Integration and Orchestration. Integration tools and API connections coordinate data flow between foundation and capability layers. Maintain flexible architecture that accommodates new tools without complete workflow rebuilds.

The API-First Philosophy

Every tool you select should have robust API capabilities enabling data exchange, workflow automation, and system replacement without manual data migration. Tools without APIs create vendor lock-in that prevents adaptation. Tools with robust APIs enable continuous evolution without technical debt accumulation.

This is one reason Clay has become central to so many AI Sales Engine implementations. Its integration with over one hundred thirty data providers through a single API layer means companies can add or swap enrichment sources without rebuilding workflows. When Vanta streamlined their RevOps, they were able to consolidate five different tools that were trying to enrich the same data point into one unified Clay workflow - eliminating redundancy while improving data quality.

The Documentation Imperative

Comprehensive documentation of data flows, integration parameters, prompt configurations, and workflow logic enables rapid adaptation when tools evolve. Investment: ten to fifteen percent of implementation time during initial setup, with ongoing maintenance as systems evolve. Companies with thorough documentation adapt to new requirements in days rather than weeks because they can precisely analyze impact and target changes.

The Skills Investment

Train teams on AI fundamentals rather than specific tool usage. A team trained on prompt engineering principles adopts new AI tools three times faster than a team trained on specific platform navigation, because they understand underlying concepts rather than just interfaces.

Skills development priorities: prompt engineering, data analysis, integration logic, performance optimization, and ethical AI usage. These transferable skills remain valuable regardless of which specific platforms dominate next year.

Maintaining Your Competitive Edge

The Continuous Innovation Cycle

Innovation isn't annual planning - it's quarterly assessment, ongoing experimentation, and systematic evolution based on market conditions.

Q1: Emerging Technology Evaluation. Every January, systematically evaluate AI capabilities that have emerged in the past twelve months. Assess strategic alignment, integration compatibility, ROI potential, and competitive advantage opportunity. Evaluate five to ten new tools, test two to three in pilot programs, adopt one to two annually based on proven results.

Q2: Competitive Intelligence Analysis. Every April, conduct comprehensive analysis of competitor AI capabilities and market positioning. Mystery shop competitor experiences.

Analyze their job postings to reveal technology priorities. Monitor their content and messaging for capability indicators. Gather customer feedback about competitor experiences. The companies that spot competitive threats early have time to respond. The ones that discover threats during lost deal reviews are already behind.

Q3: Internal Optimization Audit. Every July, analyze current system performance and identify optimization opportunities. Focus on maximizing existing capabilities before adding new technologies. Systematic analysis of existing workflows, prompt optimization, and integration efficiency often reveals twenty-plus percent improvement opportunities hiding in plain sight.

Q4: Strategic Planning and Roadmap Development. Every October, develop the AI roadmap for the following year based on technology evolution, competitive requirements, and business objectives. Maintain a three-year rolling roadmap that guides investment decisions, hiring priorities, and vendor relationships. Execute systematically rather than reactively.

The Test-and-Learn Culture

Allocate ten to fifteen percent of AI budget for experimentation with emerging capabilities. Don't wait for perfect information - test promising capabilities early while costs are low. Small-scale pilots of fifty to one hundred prospects with one to two reps over thirty to sixty day evaluation periods produce reliable data for go/no-go decisions.

The companies that discover competitive advantages before the market does are the ones testing during beta periods while everyone else is in wait-and-see mode. That's the difference between the nineteen percent and the thirty-one percent in the Gartner data.

The Partnership Strategy

Build relationships with AI vendors for early access to beta capabilities, product development influence, and market insights. Partnership benefits include early capability access, roadmap influence, preferred pricing, and industry intelligence from vendor customer bases.

The Talent Development Approach

Hire or develop team members who understand AI capabilities instinctively. Consider creating an "AI Sales Engineer" or "Head of AI Sales Operations" role responsible for evaluating emerging capabilities, managing pilot programs, and training team members. This specialized investment transforms your ability to adopt and optimize AI capabilities - it's the difference between reactive adoption and strategic evolution.

Preparing for Autonomous Sales Systems

The most significant change coming to B2B sales is the shift from AI-augmented human processes to human-supervised AI processes.

The Shift from Augmented to Autonomous

The evolution happens in measurable stages. Level 1 Automation is current reality: AI follows predetermined rules, humans make all strategic decisions. Level 2 Augmentation is current advanced implementation: AI suggests actions and provides recommendations, humans review and approve. Level 3 Autonomy is emerging: AI executes complete processes independently while humans monitor performance and handle exceptions. Level 4 Intelligence is future development: AI determines optimal strategies based on data analysis, humans set objectives and ethical guidelines.

Each level requires different organizational capabilities, different team structures, and different management approaches. The companies preparing for Level 3 and 4 now will transition smoothly. The companies waiting until these capabilities arrive will scramble to catch up.

Preparing Your Organization

This evolution isn't about replacing people - it's about elevating human roles from routine execution to strategic value creation.

The evolving sales role means less time on manual research, data entry, follow-up scheduling, and administrative tasks. More time on strategic relationship building, complex problem-solving, consultative selling, and deal orchestration. The future sales professional is a strategic coordinator who uses AI capabilities to deliver superior customer experiences, not an individual contributor who manually executes all activities.

The Skills Transition

Current core skills remain essential: relationship building, problem identification, solution presentation, negotiation, closing. But they need supplementing with emerging required skills: prompt engineering, performance analysis, system optimization, and strategic AI-human workflow coordination. And future strategic skills: pattern recognition beyond AI capabilities, emotional intelligence for relationships AI cannot handle, creative problem-solving, and ethical decision-making.

The transition works best as gradual skill addition rather than sudden role replacement. Continuous learning rather than one-time retraining. Mentorship from early adopters. Celebration of enhanced capabilities rather than fear of job displacement.

Ethical Considerations

Autonomous AI raises questions about transparency, authenticity, and consent. When should prospects know they're interacting with AI? How much disclosure is appropriate? Companies developing clear transparency policies now -

requiring disclosure when AI handles more than fifty percent of prospect interaction - find that transparency builds trust rather than reducing conversion rates. We'll explore this fully in the Bonus Chapter.

The AI-Native Sales Organization

Companies building sales operations today can design around AI from the beginning rather than adding AI to traditional processes. This creates fundamentally superior capabilities.

What AI-Native Really Means

AI-native organizations design processes, roles, and systems around AI capabilities rather than treating AI as an enhancement to existing workflows. The advantage comes from eliminating inefficiencies that traditional organizations accept as normal: manual research, administrative data entry, generic outreach, reactive follow-up, and process inconsistency.

The operating model includes a centralized AI Operations team managing prompts, optimizing integrations, and continuously improving AI performance. Distributed AI usage means every team member uses AI tools daily rather than having specialists while others use traditional methods. And a data-driven decision culture means every decision is backed by data analysis, with A/B testing as default practice.

The companies building this way are running eight hundred-plus optimizations per year compared to the traditional competitor's twelve. Compound advantage from continuous improvement creates performance gaps that become insurmountable.

The Metrics Evolution

Traditional metrics focus on activities: calls made, emails sent, meetings held. AI-augmented metrics focus on outcomes: responses generated, meetings booked, deals closed. AI-native metrics focus on intelligence: prediction accuracy, automation rate, optimization velocity, capability evolution speed, cost efficiency improvement over time.

The shift from measuring effort to measuring intelligence reflects the fundamental transformation from manual sales to systematic revenue generation.

The Organizational Structure Shift

Traditional organizations use linear handoffs: SDR qualifies, AE presents, Customer Success manages. AI-native organizations use coordinated orchestration: AI agents handle routine interactions across the entire customer lifecycle while humans focus on strategic value creation. The AI maintains complete customer context and history. Human specialists engage when expertise adds value. Customer experience remains consistent regardless of which team member is involved.

Your 12-Month Future-Proofing Roadmap

Months 1-3: Assessment and Planning. Conduct competitive position audit comparing your capabilities against market leaders, direct competitors, and emerging players. Research emerging AI capabilities, vendor roadmaps, and adoption trends. Allocate ten to fifteen percent of current AI budget for innovation and experimentation. Evaluate team capabilities and identify training needs.

Months 4-6: Pilot and Test. Select three to five promising capabilities for small-scale pilot programs. Establish clear success metrics and decision criteria before beginning. Involve pilot participants in capability evaluation. Verify integration with existing architecture before committing.

Months 7-9: Implement and Scale. Roll out proven capabilities to the full team with systematic training and documentation. Monitor performance closely during scaling to catch degradation early. Optimize based on full-scale data rather than pilot assumptions. Share success stories to maintain team enthusiasm.

Months 10-12: Optimize and Plan. Maximize performance of newly implemented capabilities through systematic optimization. Begin next assessment cycle for following year's evolution priorities. Update three-year roadmap based on current year learnings and market evolution. Share results with leadership to maintain investment commitment.

Executive Takeaways

For the CEO: Future-proofing is strategic survival, not optional enhancement. AI competitive advantages are real but temporary without continuous evolution. Companies that stop innovating after initial implementation lose advantages within twelve to eighteen months as competitors catch up and surpass. Investment required: ten to fifteen percent of AI budget allocated for innovation and continuous evolution. Expected advantage: companies with continuous innovation processes maintain twelve-plus month competitive advantages indefinitely. Organizational requirement: build culture of experimentation and learning rather than protecting current approaches. The Kodak lesson is clear - they held the future in their hands and chose to protect the present. Don't make that choice with your AI Sales Engine.

For the CRO: Your sales team's relationship with AI will evolve from tool usage to strategic partnership as capabilities advance. Future sales roles will focus on strategic activities that create more value for customers and more satisfaction for employees. Training investment must shift from tool-specific training to AI fundamentals that enable rapid adaptation to new capabilities. Competitive monitoring should include AI capability assessment of competitors, not just messaging and pricing analysis. Sales professionals who develop AI coordination skills will be dramatically more valuable than those who resist the transition.

For the CMO: Marketing and sales AI capabilities will converge, requiring integrated strategy and shared intelligence platforms. Content creation, campaign optimization, and personalization will become increasingly AI-driven, requiring marketing teams to focus on strategy rather than execution. First-mover advantage in AI adoption creates compounding benefits that are difficult for competitors to overcome. Marketing measurement will evolve from attribution to prediction, enabling proactive rather than reactive campaign management.

For the CFO: AI investment ROI increases over time as capabilities compound and operational efficiency improves continuously. Future-proofing investment of ten to fifteen percent of AI budget prevents costly catch-up spending when competitive gaps emerge. Valuation impact: companies demonstrating systematic AI evolution command premium multiples compared to static implementations. Risk mitigation: companies without AI evolution strategy face increasing competitive vulnerability as market capabilities advance rapidly.

For the CTO/RevOps: Technical architecture must prioritize flexibility and integration capability over current feature optimization. API-first tool selection enables rapid capability evolution without system rebuilds. Documentation investment provides compound returns as systems evolve and new capabilities are integrated. Team skill development should focus on transferable AI fundamentals rather than platform-specific training. Modular architecture enables individual capability upgrades without disrupting entire system.

The question isn't whether your AI Sales Engine works. It does. The question is whether you'll protect the present or build for the future. The Kodak story proves that holding the future in your hands means nothing if you choose not to develop it. Build the organizational muscle to continuously evolve, adopt, and stay ahead of a technology curve that never stops moving.

In the Bonus Chapter, we'll address the ethical dimensions of AI-powered sales - the transparency, privacy, and responsibility frameworks that ensure your AI Sales Engine builds trust rather than destroying it.

BONUS CHAPTER:

AI Ethics & Responsible Sales Enablement

I need to tell you something about this book.

When I started writing it, I did what this entire book teaches you to do - I used AI as a force multiplier. I fed it my frameworks, my client experiences, my methodologies, and asked it to help me organize twenty years of sales strategy into something coherent and actionable. The AI was incredible. It worked fast, it structured beautifully, and it produced content that sounded exactly like the kind of book I wanted to write.

There was just one problem: the first pass invented thirteen case studies that never happened.

They were convincing. A $47 million SaaS company that called me on a Tuesday. A $34 million cybersecurity firm where four executives gave four different answers in a boardroom. A $52 million fintech CTO who forwarded me a spreadsheet with forty-seven vendors. Every one of them had specific revenue figures, realistic timelines, believable dialogue, and results that mapped perfectly to the points I was making. They read like real stories. They felt like real stories. Some of them were so well-crafted that I had to stop and ask myself whether they actually happened.

They didn't.

The AI had done what AI does - it generated plausible, persuasive content that fit the pattern it was asked to follow. It wasn't lying in the way a person lies. It was filling gaps with the most statistically likely version of reality. And if I hadn't caught it - if I'd published those stories as written - I would have built an entire book about trust and systems on a foundation of fabricated evidence.

That realization changed how I finished this book. We rebuilt every chapter opening from real experiences, real conversations, and real principles I could stand behind. It took longer. It was harder. But this is a book about building revenue engines that earn trust and generate results - and I couldn't ask you to hold your AI to a standard I wasn't willing to hold myself to.

Now here's why this story matters for your business: if AI can fool the author of a book about AI, it can fool your sales team.

Right now, your enrichment engine is generating personalization hooks for every prospect on your Top 25 list. Your reps are taking those hooks and dropping them into emails. And most of the time, the data is accurate and the personalization is genuine. But some percentage of the time - and it might be higher than you think - the AI is fabricating. It's referencing LinkedIn posts that don't exist. It's congratulating prospects on funding rounds that happened two years ago. It's mentioning a "recent expansion" into a market the company just exited.

Your reps don't catch it because the content sounds right. The prospect notices it because they know their own company. And in that moment, you haven't just sent a bad email - you've told a potential customer that your company is willing to fake its homework. In B2B, where every deal is built on trust and credibility, that's not a recoverable mistake. That's a relationship you'll never get back.

Efficiency doesn't equal truth. Speed doesn't equal accuracy. And the same AI capabilities that make your sales team ten times more productive can make them ten times more dangerous if nobody is checking the work.

At the end of the day, B2B sales is still about trust. It's still about looking someone in the eye, shaking their hand, and knowing that what you told them is real. AI can make you faster. It can make you smarter. It can

put you in front of the right person at the right time with the right message. But it cannot replace the moment where a buyer decides they believe you. That's human. That's earned. And it will be a long time before any algorithm takes that away.

This chapter is about making sure your AI never puts that moment at risk - the verification systems, compliance frameworks, and quality controls that keep your sales engine honest so that when the handshake happens, it means something.

The Five Pillars of Ethical AI Sales

Ethical AI sales isn't about limiting capabilities - it's about channeling capabilities toward sustainable competitive advantages rather than short-term tricks that create long-term problems.

Pillar 1: Transparency About AI Usage

The question isn't whether to use AI in sales - it's whether to be honest about using AI in sales. Transparency builds trust faster than deception creates engagement.

AI-generated content requires disclosure when prospects ask directly or when content accuracy depends on understanding AI involvement. Autonomous agents handling initial interactions should identify themselves as AI systems with human oversight available. Predictive analytics influencing sales approaches should be explained when they affect prospect experience.

Here's what's counterintuitive: companies that implement upfront AI disclosure consistently see response rates increase rather than decrease. An email opening like "Our AI system analyzed your company's recent growth signals and thought we should connect about this specific opportunity" outperforms "I was researching companies like yours and noticed this generic observation." Authenticity builds trust faster than manufactured personalization creates engagement. Prospects appreciate honesty about data sources and research methods. AI disclosure positions the company as innovative rather than deceptive.

The transparency framework is straightforward. Be honest about data sources: "Using publicly available information about your company's expansion" rather than hiding that you've researched their public activities. Disclose automation appropriately: "Our AI system scheduled this meeting based on your calendar availability" rather than pretending a human manually coordinated scheduling. Acknowledge AI limitations: "AI analysis suggests your team might benefit from this" rather than claiming certainty based on limited data. And always provide human escalation: "Want to speak directly with a person?" should be available and honored immediately.

The principle: prospects should never feel deceived when they discover AI involvement, because they were informed from the beginning.

Pillar 2: Data Privacy and Protection

Data ethics in AI sales requires clear boundaries about what information can be collected, how it can be used, and how long it can be retained.

Public data from LinkedIn profiles, company websites, and press releases is generally acceptable because prospects have chosen to make this information publicly available. Purchased data from vendors like ZoomInfo or Clearbit is acceptable if vendors comply with privacy regulations and prospects had reasonable opportunity to consent or opt out. Scraped data from web harvesting exists in legal gray area with significant ethical concerns. Personal data including health information, financial details, or political affiliations is highly restricted and usually inappropriate for sales prospecting.

Companies that implement a "public data only" policy typically lose a small percentage of potential data points but gain substantially higher close rates because prospects trust their approach to data handling. When prospects ask about data sources and you can honestly respond that you use only information they've chosen to make publicly available, that transparency becomes a competitive differentiator - especially when competitors can't make the same claim.

The regulatory landscape makes this increasingly urgent. Total GDPR fines in 2024 surpassed five billion euros, with enforcement actions reaching well beyond the technology sector into finance, healthcare, and energy. LinkedIn received a three hundred ten million euro fine from the Irish Data Protection Commission for misuse of user data for behavioral analysis and targeted advertising. Clearview AI was fined thirty and a half million euros by the Dutch Data Protection Authority for illegal facial image collection. These aren't theoretical risks - they're real financial consequences that affect companies of every size.

Data handling requirements are non-negotiable. Consent: do prospects understand that you have their information and how you obtained it? Storage: all prospect data must be encrypted, access-controlled, and secured with audit trails. Retention: delete data when no longer needed for business purposes and when prospects explicitly opt out. Sharing: never sell prospect data or share with third parties without explicit consent.

Pillar 3: Accuracy and Truthfulness

AI systems can generate convincing but completely false information. Human oversight is essential to prevent AI hallucinations from damaging prospect relationships and company reputation.

This isn't theoretical. As I described in this chapter's opening, our AI fabricated thirteen case studies in the first draft of this book. They were so convincing that I had to stop and ask myself whether they actually happened. They didn't. If AI can fool the author of a book about AI, it can fool your sales team.

The problem extends to every AI-generated personalization hook your reps send. Some percentage of the time - and it might be higher than you think - the AI is referencing LinkedIn posts that don't exist, congratulating prospects on funding rounds that happened two years ago, or mentioning "recent expansion" into markets the company just exited. Your reps don't catch it because the content sounds right. The prospect notices because they know their own company. And in that moment, you've told a potential customer that your company is willing to fake its homework.

Human validation requirements should be absolute: never send AI-generated content without human review, especially when content includes specific facts about prospect companies. Verify facts by checking original sources rather than trusting AI summaries. Flag uncertain information for additional research rather than including potentially false details. Train team members to recognize hallucinations by understanding what types of errors AI commonly makes.

A simple "two-check" policy works: AI generates personalization content, sales rep validates accuracy against original sources, manager spot-checks random samples for quality control. The time investment is modest - three to five minutes per prospect for validation - but the reputation protection is invaluable.

When AI does get it wrong despite best efforts, ethical response actually strengthens relationships. Acknowledge mistakes immediately. Apologize genuinely. Correct the record with accurate information. Explain prevention measures. I've seen relationships that started with an AI error convert to significant contracts specifically because the rep's ethical response demonstrated trustworthiness. Honesty about mistakes builds more trust than pretending mistakes don't happen.

Pillar 4: Fairness and Non-Discrimination

AI systems can perpetuate or amplify biases present in training data, leading to discriminatory outcomes that are both unethical and illegal.

This is a documented, widespread problem. Gartner research found that fifty-three percent of HR leaders are concerned about potential bias and discrimination from AI, and nearly sixty percent of employees believe AI is making bias worse rather than better. If your AI system learns from historical sales data when certain groups were systematically underrepresented or undervalued, it may continue those patterns automatically.

The bias might be subtle. A lead scoring system trained on historical data from an era when women were underrepresented in executive roles might consistently rate female executives lower than men with identical qualifications. The bias is unintentional but the impact is real: affected prospects receive less attention from sales teams, creating a self-fulfilling prophecy that reinforces the original bias.

Bias prevention requires regular algorithm audits analyzing patterns in AI recommendations. Diverse training data representing all customer segments you want to serve. Human oversight for high-stakes decisions. And transparent criteria for scoring and prioritization that allow team members to recognize potential unfairness.

Companies that conduct bias audits and broaden their ICP definitions to eliminate discriminatory patterns often discover they've been leaving significant market opportunity on the table. Correcting bias doesn't just make you ethical - it frequently increases your addressable market and improves product-market fit by serving more diverse customer segments.

Pillar 5: Human Accountability

AI systems should augment human judgment, not replace human responsibility. High-stakes decisions require human approval and oversight.

The most expensive failures happen when companies fully automate prospect qualification without human review. An AI system configured to automatically reject low-scoring prospects will eventually reject someone who would have become your largest customer - because the company had unusual characteristics that the AI's pattern-matching couldn't interpret correctly. AI excels at pattern recognition but struggles with exceptional cases requiring context, nuance, and creative problem-solving.

Human approval should be required for major decisions: communications that make commitments, pricing proposals, strategic account decisions, and anything that could significantly affect customer relationships. AI recommendations should always include human override capability. Clear escalation paths must exist for situations where AI approaches seem inappropriate. And regular review of AI decisions should identify patterns that enable continuous improvement.

The key principle: AI provides data and suggestions, but humans maintain responsibility for decisions and their consequences.

Building Your Ethical AI Framework

The Ethical AI Policy Document

Comprehensive written policy creates clarity about acceptable practices and provides guidance for ambiguous situations. Include acceptable use cases for AI with specific examples, prohibited applications like fabricating information or unauthorized data collection, data collection and privacy standards, transparency and disclosure requirements, accuracy verification procedures, bias prevention measures, and accountability structure.

Consider publishing your AI ethics policy publicly. Companies that do consistently find it becomes a competitive differentiator that attracts trust-conscious customers and demonstrates thought leadership.

Implementation Approach

Step 1: Leadership Alignment. CEO, CRO, CMO, and Legal counsel agree on ethical principles and commit resources. Board review ensures the framework won't be compromised during pressure to meet short-term targets.

Step 2: Policy Development. Draft comprehensive policy with legal review. Test against real scenarios by reviewing current AI practices. Refine based on practical considerations - the policy should provide clear guidance without being so restrictive that it prevents effective AI usage.

Step 3: Team Training. Comprehensive ethics training for all team members who use AI. Scenario-based learning using realistic situations they're likely to encounter. Certification requirement before AI tool access ensures understanding and competence. The difference between trained and untrained teams is dramatic - formal ethics training typically improves compliance rates from the sixty to seventy percent range up to ninety-five percent.

Step 4: Monitoring and Enforcement. Random audit of AI-generated content. Ethics reporting channel for concerns or questions without fear of retaliation. Clear consequences for violations consistently applied regardless of seniority. Regular policy updates as technology evolves.

The Ethics Committee

Cross-functional committee with sales leadership, legal counsel, technical leadership, and ideally an external customer representative. Responsibilities include reviewing ethical dilemmas and edge cases, updating policies as technology evolves, investigating potential violations, and providing ongoing guidance.

Quarterly committee reviews consistently identify potential ethical issues before they become customer-facing problems. The committee creates systematic oversight without bureaucracy and provides a forum for addressing questions before they become crises.

Common Ethical Dilemmas and Solutions

The Personalization vs. Privacy Balance

Deep personalization requires extensive data, but comprehensive collection can feel intrusive. The solution: limit collection to information prospects have chosen to make publicly available. Focus personalization on company information - growth, technology, market position - rather than personal information like hobbies, family, or political views. Company-based personalization actually generates higher response rates than personal-based personalization, so you're not sacrificing effectiveness for ethics.

The Speed vs. Accuracy Tradeoff

Autonomous AI agents can respond in minutes, but fast responses sometimes include inaccurate information. The solution: use autonomous agents for routine qualification and scheduling where accuracy risks are low. Require human review before agents make complex statements about product capabilities, pricing, or prospect-specific recommendations.

Include clear disclaimers in autonomous communications and always offer human escalation.

The AI Attribution Question

Should prospects know when AI generated their email content or scheduled their meeting? The solution: don't claim AI-generated content as human-written, but don't unnecessarily emphasize AI involvement when it's not relevant. Respond honestly when asked directly. Focus communication on value delivered rather than methods. Companies that develop standard transparent responses about AI usage consistently report no negative reactions and often generate productive conversations about AI innovation.

The Competitive Intelligence Ethics

AI can gather extensive competitive intelligence, but some methods raise ethical questions. The solution: limit competitive intelligence to publicly available information. Avoid creating fake personas for mystery shopping, recording competitor calls, or using deceptive methods for confidential information. Companies that maintain ethical competitive intelligence practices often find that their restraint becomes a trust-building advantage when prospects discover competitors use questionable research methods.

The Predictive Scoring Bias

AI scoring systems might systematically disadvantage certain prospect types based on historical training bias. The solution: quarterly analysis of scoring patterns to identify potential discrimination. Test systems for unfair treatment based on company size, industry, geography, or leadership demographics. Adjust algorithms when bias is detected using diverse, balanced training data.

Regulatory Compliance

GDPR for European Prospects

European prospects have specific rights regardless of where your company is located. Lawful basis is required for processing personal data. Individual rights include access, deletion, correction, and data portability. Consent requirements apply for certain processing types. Data breach notification must reach supervisory authorities within seventy-two hours. Privacy by design principles require protection built into systems from the beginning.

Practical implementation: establish data processing agreements with all vendors handling European data. Update privacy policy to explicitly mention AI usage. Create mechanisms for handling deletion requests across all systems. Maintain audit trails of processing activities.

GDPR applies to any company processing European personal data regardless of location. The enforcement data from 2024 makes this clear - fines totaling over five billion euros, with the LinkedIn, Clearview AI, and Enel Energia actions demonstrating that regulators are actively enforcing across sectors and geographies.

CCPA for California Prospects

Key requirements include disclosure of data collection practices, right to opt-out of data sales, right to deletion, and non-discrimination provisions. Most companies find it simpler to implement CCPA compliance nationally rather than identifying California prospects specifically. National compliance becomes a marketing advantage: "We respect your privacy everywhere, not just where the law requires it."

SOC-II for Enterprise Customers

Enterprise customers increasingly require SOC-II certification from vendors handling their data. Requirements include security controls, availability and processing integrity, confidentiality protections, privacy safeguards, and annual certified audit. Companies without SOC-II certification lose enterprise deals to certified competitors regardless of product quality. This is increasingly table stakes for enterprise B2B sales.

Industry-Specific Regulations

Healthcare organizations face HIPAA requirements that affect how they can be approached. Focus organizational data - hospital expansion, technology investments, compliance challenges - rather than anything that could be construed as patient information.

Financial services companies must retain AI-generated communications for compliance auditing under SOX and FINRA requirements. Automated archiving systems that capture all AI-generated content enable compliance while maintaining efficiency.

Executive Takeaways

For the CEO: Ethics isn't compliance overhead - it's competitive advantage and brand protection. Companies implementing ethical AI frameworks consistently discover that responsible practices improve rather than limit business results. Customer trust is your most valuable asset. Ethical AI practices protect and enhance that trust rather than trading it for short-term efficiency. The companies that establish ethical AI frameworks now create market leadership positions that become harder for competitors to replicate over time. Invest in ethics infrastructure as seriously as you invest in sales technology.

For the CRO: Sales team ethics training prevents costly mistakes that damage customer relationships and company reputation. Implement verification processes that ensure AI accuracy without creating bureaucratic bottlenecks. Create culture where team members feel comfortable raising ethical concerns without fear of retaliation. Ethical selling builds stronger, more profitable long-term relationships than aggressive approaches that prioritize volume.

For the CMO: Brand reputation is more valuable than any single campaign result. AI ethics policy should be part of brand strategy, not just compliance requirement. Consider publishing AI ethics commitments publicly as competitive differentiator. Monitor marketing practices for ethical compliance as rigorously as you monitor performance metrics.

For the CFO: Regulatory compliance investment prevents fines that can reach hundreds of millions under GDPR and significant amounts under CCPA. Total GDPR fines exceeded five billion euros in 2024 alone. Ethical practices reduce customer churn by building stronger trust relationships. Insurance against reputational damage is cheaper than recovering from reputational crisis. Budget for ethics infrastructure including training, monitoring, and compliance activities.

For the CTO/RevOps: Data architecture must support privacy requirements from design, not as an afterthought. Implement comprehensive data governance including encryption, access controls, and audit trails. Regular bias testing of AI systems prevents discriminatory outcomes that create legal liability. Documentation of AI decision processes enables compliance auditing and continuous improvement. Build systems that can respond to deletion requests and data access requests efficiently.

This chapter started with a confession about the fake case studies AI generated in the first draft of this book. That experience taught me something I want to leave you with: the same AI capabilities that make your sales team dramatically more productive can make them dramatically more dangerous if nobody is checking the work. Efficiency doesn't equal truth. Speed doesn't equal accuracy. And no amount of pipeline generation is worth losing the trust that every B2B relationship is built on.

Build the verification systems. Implement the compliance frameworks. Train your team on what ethical AI usage actually looks like. Not because regulators require it - though they increasingly do - but because your customers deserve it, your team members deserve it, and your company's long-term success depends on it.

At the end of the day, B2B sales is still about trust. AI can make you faster, smarter, and more precise. But it cannot replace the moment where a buyer decides they believe you. That's human. That's earned. And it's worth protecting.

Resource Library

Recommended AI Sales Platforms

- Detailed platform comparisons and selection criteria with
- Implementation complexity and resource requirement analysis with
- Pricing models and total cost of ownership calculations with ROI
- Integration capabilities and technical specifications with

Industry Benchmarks and Performance Standards

- Performance benchmarks by industry and company size with statistical
- Success criteria and measurement frameworks with baseline
- Competitive analysis and market positioning with differentiation
- Best practice examples and case study library with implementation

References & Further Reading

Primary Research Sources

Forrester Research. (2024). *The state of B2B sales technology: AI adoption and competitive advantage.* Forrester B2B Sales Research Division.

Gartner. (2024). *Magic quadrant for sales force automation platforms.* Gartner Sales Technology Research.

Grand View Research. (2024). *Sales enablement platform market size, share and forecast 2030.* https://www.grandviewresearch.com/industry-analysis/sales-enablement-platform-market-report

HubSpot Research. (2024). *2024 AI trends for sales: Adoption, performance, and future predictions.* HubSpot Sales Research Division. https://offers.hubspot.com/ai-sales

McKinsey Global Institute. (2024). *The next frontier: AI adoption and competitive advantage in B2B sales.* McKinsey & Company.

SaaS Capital. (2025). *AI adoption among private SaaS companies and its impacts on spending and profitability.* https://www.saas-capital.com/blog-posts/ai-adoption-among-private-saas-companies-and-its-impacts-on-spending-and-profitability/

Salesforce Research. (2024). *State of sales: 9th edition research report.* Salesforce Sales Cloud Research Division. https://www.salesforce.com/resources/research-reports/state-of-sales/

Technology and Implementation Guides

6sense. (2024). *Revenue AI benchmark report: B2B buyer behavior and predictive intelligence trends.* 6sense Revenue AI Research.

Bombora. (2024). *The state of intent data: 2024 benchmark report.* Bombora Company Surge Research.

Gong.io. (2024). *Conversation intelligence best practices: Implementation and optimization guide.* Gong Revenue Intelligence.

Salesforce. (2024). *Integration architecture best practices guide.* Salesforce Developer Documentation. https://developer.salesforce.com/docs/atlas.en-us.integration_patterns_and_practices.meta/integration_patterns_and_practices/

ZoomInfo. (2024). *The state of pipeline generation: Data-driven prospecting benchmark study*. ZoomInfo Research Division.

Regulatory and Compliance Resources

European Union. (2024). *General Data Protection Regulation (GDPR): AI-specific*

European Union. (2024). *General Data Protection Regulation (GDPR): AI-specific guidance and implementation requirements*. European Data Protection Board.

Federal Trade Commission. (2024). *Artificial intelligence and algorithms: Business guidance and compliance requirements*. FTC Bureau of Consumer Protection.

International Organization for Standardization. (2024). *ISO/IEC 23053: Framework for AI systems using machine learning*. ISO/IEC JTC 1/SC 42.

Industry Analysis and Market Research

CFO Research. (2024). *AI ROI measurement and attribution: Best practices for B2B technology companies*. CFO Research Services.

Forrester Research. (2024). *The total economic impact of AI-powered sales technology*. Forrester Consulting Economic Impact Study.

G2 Research. (2024). *AI sales software market report: Vendor analysis and buyer behavior trends*. G2 Research Grid.

Manufacturing Leadership Council. (2024). *Digital transformation in industrial sales: 2024 benchmark study*. Manufacturing Leadership Council.

MuleSoft. (2024). *API integration patterns and best practices*. MuleSoft Documentation. https://docs.mulesoft.com/general/api-led-overview

Sales Enablement Society. (2024). *Best practices in sales process documentation and training.* Sales Enablement Society Research.

285

About **Doug Foley**

Doug Foley grew up the youngest of eight children in an entrepreneurial family. At nine years old, he spent Saturday mornings at his father's wood shop, dusting machinery and sweeping floors before they'd head off to play golf. It wasn't until years later that his dad explained the lesson: "The reason I wanted you to do that job was so you learned the value of hard work, so you would work hard enough to never want to do that job again."

That lesson landed harder than intended. In Doug's senior year of high school, his father was forced to close the business after losing his largest customer. Watching his hero struggle planted something that would shape his entire career --- a quiet obsession with understanding how companies find and keep customers, so the people who build them never have to face that moment unprepared.

The obsession led Doug to study marketing, then into nearly two decades of building the skills to back it up. He sold online courses. Managed sales teams. Launched tech startups. Started businesses, some of which he still runs today. Along the way, he interviewed over 150 entrepreneurs on his podcast *The Happiness of Pursuit*, wrote *Breakout Blueprint*, and delivered a 2021 TEDx Talk called "How Entrepreneurs Inspire Change" --- all in pursuit of a single question: what separates the companies that scale from the ones that stall?

The answer turned out to be systems. The entrepreneurs who achieved the largest exits weren't the most passionate or visionary. They were the most systematic. That insight led to the Customer Loyalty Framework and eventually to the AI Sales Engine methodology --- the approach detailed in this book.

Today, Doug is the CEO of FoleyMedia.ai and host of the *9 Figure Exits* show, where entrepreneurs, operators and investors share boardroom playbooks for building, scaling, and selling companies. He partners with founder-CEOs, private equity firms, and enterprise leadership teams to design AI-powered revenue systems that shorten sales cycles, grow pipeline, and increase enterprise value. His work spans growth-stage companies seeking their first systematic sales motion to established organizations preparing for nine-figure exits.

Doug lives in Virginia Beach with his wife --- the same woman his father handed a silver spoon to on their wedding day with the words, "He's yours now."

Connect with Doug LinkedIn: linkedin.com/in/douglasjfoley

Company: foleymedia.ai

Podcast: 9 Figure Exits

Your AI Transformation Begins with a Single Decision

Will you lead your market or follow your competitors? Will you build a systematic revenue engine or continue hoping for growth? Will you harness AI as a strategic capability or treat it as just another tool?

The companies implementing AI Sales Engines today are defining the competitive landscape tomorrow. The window for first-mover advantage is closing rapidly, but systematic implementation still provides 12-18 months of sustainable competitive advantage.

What's Next

If you've made it this far, you've done more than read a book---you've absorbed a system. The frameworks, workflows, and architectures in these pages aren't theoretical. They're running inside real companies right now, generating pipeline, shortening cycles, and changing how leadership teams think about revenue.

But reading and implementing are two different things.

The gap between understanding the AI Sales Engine methodology and actually building one inside your organization is where most companies stall. Not because the concepts are difficult, but because execution requires someone who has done it before---someone who knows which shortcuts save months and which ones create expensive problems six quarters later.

That's what FoleyMedia.ai does. Doug and his team work directly with founder-CEOs, private equity operating partners, and enterprise leadership teams to design, build, and activate AI-powered revenue systems. The engagement typically starts with a 90-day architecture sprint---the same prove-then-scale approach described in this book---and evolves into an ongoing partnership as the system matures.

If you're running a B2B company with real revenue, real ambition, and a leadership team ready to move, the next step is a conversation.

foleymedia.ai

About This Index: This comprehensive index provides page references for all major concepts, frameworks, methodologies, and tools discussed throughout "AI Sales Engine: How CEOs Architect, Deploy & Scale AI-Powered Revenue Systems." Use this index to quickly locate specific information and cross-reference related concepts across chapters.